Xiaoxia Huang

Portfolio Analysis

Studies in Fuzziness and Soft Computing, Volume 250

Editor-in-Chief

Prof. Janusz Kacprzyk
Systems Research Institute
Polish Academy of Sciences
ul. Newelska 6
01-447 Warsaw
Poland
E-mail: kacprzyk@ibspan.waw.pl

Further volumes of this series can be found on our homepage: springer.com

Vol. 233. Cengiz Kahraman (Ed.)
Fuzzy Engineering Economics with Applications, 2008
ISBN 978-3-540-70809-4

Vol. 234. Eyal Kolman, Michael Margaliot
Knowledge-Based Neurocomputing: A Fuzzy Logic Approach, 2009
ISBN 978-3-540-88076-9

Vol. 235. Kofi Kissi Dompere
Fuzzy Rationality, 2009
ISBN 978-3-540-88082-0

Vol. 236. Kofi Kissi Dompere
Epistemic Foundations of Fuzziness, 2009
ISBN 978-3-540-88084-4

Vol. 237. Kofi Kissi Dompere
Fuzziness and Approximate Reasoning, 2009
ISBN 978-3-540-88086-8

Vol. 238. Atanu Sengupta, Tapan Kumar Pal
Fuzzy Preference Ordering of Interval Numbers in Decision Problems, 2009
ISBN 978-3-540-89914-3

Vol. 239. Baoding Liu
Theory and Practice of Uncertain Programming, 2009
ISBN 978-3-540-89483-4

Vol. 240. Asli Celikyilmaz, I. Burhan Türksen
Modeling Uncertainty with Fuzzy Logic, 2009
ISBN 978-3-540-89923-5

Vol. 241. Jacek Kluska
Analytical Methods in Fuzzy Modeling and Control, 2009
ISBN 978-3-540-89926-6

Vol. 242. Yaochu Jin, Lipo Wang
Fuzzy Systems in Bioinformatics and Computational Biology, 2009
ISBN 978-3-540-89967-9

Vol. 243. Rudolf Seising (Ed.)
Views on Fuzzy Sets and Systems from Different Perspectives, 2009
ISBN 978-3-540-93801-9

Vol. 244. Xiaodong Liu and Witold Pedrycz
Axiomatic Fuzzy Set Theory and Its Applications, 2009
ISBN 978-3-642-00401-8

Vol. 245. Xuzhu Wang, Da Ruan, Etienne E. Kerre
Mathematics of Fuzziness – Basic Issues, 2009
ISBN 978-3-540-78310-7

Vol. 246. Piedad Brox, Iluminada Castillo, Santiago Sánchez Solano
Fuzzy Logic-Based Algorithms for Video De-Interlacing, 2010
ISBN 978-3-642-10694-1

Vol. 247. Michael Glykas
Fuzzy Cognitive Maps, 2010
ISBN 978-3-642-03219-6

Vol. 248. Bing-Yuan Cao
Optimal Models and Methods with Fuzzy Quantities, 2010
ISBN 978-3-642-10710-8

Vol. 249. Bernadette Bouchon-Meunier, Luis Magdalena, Manuel Ojeda-Aciego, José-Luis Verdegay, Ronald R. Yager (Eds.)
Foundations of Reasoning under Uncertainty, 2010
ISBN 978-3-642-10726-9

Vol. 250. Xiaoxia Huang
Portfolio Analysis, 2010
ISBN 978-3-642-11213-3

Xiaoxia Huang

Portfolio Analysis

From Probabilistic to Credibilistic and Uncertain Approaches

Author

Xiaoxia Huang
School of Economics and Management
University of Science and
Technology Beijing
Beijing 100083, China
E-mail: hxiaoxia@manage.ustb.edu.cn

ISBN 978-3-642-26249-4 e-ISBN 978-3-642-11214-0

DOI 10.1007/978-3-642-11214-0

Studies in Fuzziness and Soft Computing ISSN 1434-9922

Typeset & Cover Design: Scientific Publishing Services Pvt. Ltd., Chennai, India.

Printed in acid-free paper

9 8 7 6 5 4 3 2 1

springer.com

Preface

The most salient feature of security returns is uncertainty. The purpose of the book is to provide systematically a quantitative method for analyzing return and risk of a portfolio investment in different kinds of uncertainty and present the ways for striking a balance between investment return and risk such that an optimal portfolio can be obtained.

In classical portfolio theory, security returns were assumed to be random variables, and probability theory was the main mathematical tool for handling uncertainty in the past. However, the world is complex and uncertainty is varied. Randomness is not the only type of uncertainty in reality, especially when human factors are included. Security market, one of the most complex markets in the world, contains almost all kinds of uncertainty. The security returns are sensitive to various factors including economic, social, political and very importantly, people's psychological factors. Therefore, other than strict probability method, scholars have proposed some other approaches including imprecise probability, possibility, and interval set methods, etc., to deal with uncertainty in portfolio selection since 1990's. In this book, we want to add to the tools existing in science some new and unorthodox approaches for analyzing uncertainty of portfolio returns. When security returns are fuzzy, we use credibility which has self-duality property as the basic measure and employ credibility theory to help make selection decision such that the decision result will be consistent with the laws of contradiction and excluded middle. Being aware that one tool is not enough for solving complex practical problems, we further employ uncertain measure and uncertainty theory to help select an optimal portfolio when security returns behave neither randomly nor fuzzily.

One core of portfolio selection is to find a quantitative risk definition of a portfolio investment. Another interesting feature of the book is that it introduces a new risk definition, i.e., risk curve, besides already known risk definitions of variance, semivariance, and probability of a disastrous loss level. Risk curve describes each likely loss level and the corresponding occurrence chance of each loss. So it is instinct and safe for investors to use risk curve to control their risk. Furthermore, the book provides the extensions of the risk definitions to other types of uncertainty other than randomness.

This book consists of 5 chapters. Chapter 1 introduces general principles upon which portfolio selection problem is analyzed. Chapter 2 provides a variety of models with numerous application examples for portfolio selection with random returns. Risk curve is introduced and models based on risk curve are provided in this chapter. For better understanding of the selection ideas in random environment, fundamentals of probability theory are reviewed at the beginning of Chapter 2. Chapter 3 starts with an introduction of fundamentals of credibility theory concerning fuzzy portfolio selection and then introduces a spectrum of credibilistic portfolio selection models including mean-risk model, β-return-risk model, credibility minimization model, mean-variance model, mean-semivariance model, and entropy optimization model. Crisp equivalents of the credibilistic models are given when security returns are triangular fuzzy variables, trapezoidal fuzzy variables, normal fuzzy variables and equipossible fuzzy variables. A hybrid intelligent algorithm is also presented for solution of the credibilistic models in general cases. Chapter 4 first offers necessary knowledge about uncertainty theory which will be used in portfolio selection with neither random nor fuzzy uncertain returns. Then a series of uncertain selection models are provided and the crisp equivalents are presented. Chapter 5 offers extensions of the basic portfolio selection models such that the optimal portfolio can be dispersed enough to a required extent.

The book provides a systematic, self-contained, and up-to-date portfolio analysis method. With numerous examples and necessary remarks, it is quite readable. The book is interesting because it introduces some new quantitative risk definitions and adds to the existing tools and techniques some additional apparatus for investment optimization which will be powerful in many specific cases. It is suitable for researchers and students who are interested in the fields of portfolio selection as well as capital budgeting, investment optimization, and risk analysis, etc.

I would like to thank my parents, colleagues, friends and family members who encouraged and helped me to finish this work. I would also like to thank my graduate students Qiming Pan, Wenying Shen and Wenjing Gao who made a number of corrections. This work was supported by National Natural Science Foundation of China Grant No. 70871011 and New Century Excellent Talents in University. I owe thanks to their financial support. Finally, I express my deep gratitude to Professor Janusz Kacprzyk for his valuable comments and suggestions on the book and his generosity to allow me to publish the book in his series.

October 2009 Xiaoxia Huang

Contents

Chapter 1
What Is Portfolio Analysis

A portfolio is a combination of a number of securities. Portfolio analysis is a quantitative method for selecting an optimal portfolio that can strike a balance between maximizing the return and minimizing the risk in various uncertain environments. To select the optimal portfolio, we must first answer the questions "what is return of a portfolio" and "what is risk of the portfolio". If we could only use the natural language like "the likely gain of the portfolio" to describe return and "the likely loss of the portfolio" to describe risk, we would not be able to quantify return and risk of the portfolio. Then it would be impossible to compare the return level and risk level of portfolios, let alone find the maximum return and minimum risk. To use measurable terms to define return and risk, we should start with input data, i.e., the individual security returns.

1.1 Security Return

Individual security returns are the basic information for the investors. Every decision is made based on this information. The security return is expressed by the rate of return which is defined as

$$\frac{\text{Receipt-Expenditurn}}{\text{Expenditure}}.$$

Without considering transaction cost, tax factors and stock split, the rate of return can also be defined as

$$\frac{\text{Ending price of a security-Beginning price+Cash dividend}}{\text{Beginning price}}.$$

For example, the return of a security in 2007 is

$$\frac{\text{(Closing price, 2007)-(Closing price, 2006)+(Dividend, 2007)}}{\text{(Closing price, 2006)}}.$$

X. Huang: Portfolio Analysis: From Probab. to Credibilistic, STUDFUZZ 250, pp. 1–9.
springerlink.com

Similarly, the return of a security in February, 2007 is

$$\frac{\text{(Closing price, Feb. 2007)-(Closing price, Jan. 2007)+(Dividend, Feb. 2007)}}{\text{(Closing price, Jan. 2007)}}.$$

The most salient feature of the security return is uncertainty. Sometimes it is high, and sometimes it is low. Sometimes it is positive, and sometimes it may be negative. Therefore, we can not use a deterministic number to describe it. It is reasonable that a variable should be used for description.

Random variable is the earliest employed variable to describe the security return and is still widely used nowadays. When investors say that the return of Security A will be 0.08 with probability 25%, 0.10 with probability 50% and 0.12 with probability 25%, they are using random variable to describe the security return. Like playing dice, the investors believe that three outcomes, i.e., the return being 0.08, being 0.10 and being 0.12, will appear randomly and that the occurrence chances will be 25%, 50% and 25%, respectively.

By using random variable, an assumption should be satisfied that the historical data of a security return are able to reflect the future return of the security. However, this assumption cannot always be met. It is usually agreed that when evaluating a security return, the investors should consider three types of factors, i.e., general economic factor, industry factor and the company factor [81]. General economic factor refers to those policies and events that influence the macro-economic growth. It includes the fiscal policy, monetary policy, inflation, international monetary devaluation, political events in the country, etc. Industry factor refers to those measures that influence an industry to prosper or suffer in the long run or during the expected near-term economic environment. Examples include import or export quotas or taxes, excess supply or shortage of a resource, or government-imposed regulations on an industry, etc. Company factor refers to the past performances and future prospect of the company whose security is listed on the security market. None of these factors influence the security return randomly. They affect the security return through people's psychology. Uncertainty of the security return remains even when all the factors are known. In addition, the security market is usually very sensitive. Other non-economic factors like the success of an invention, an accident or even a hard-to-verify message may influence the security return. In short, human factor contributes to uncertainty of the security return. Thus, in some situations, the investors may not think that the past data of the security return can well reflect the future return of the security. They may like to use experts' knowledge and their own experience to evaluate the future return of the security. The prediction is usually expressed in the fuzzy form like "The rate of return of security A is around 0.12". Like when we guess a man's age, we say he is "about 30" instead of "probably 30". The evaluation of the security return contains much subjectivity and is of fuzziness rather than randomness. In this situation, *fuzzy variable* can be employed to reflect the fuzziness of the security return.

So far, we believe that security returns may behave randomly or fuzzily. But in fact, there are cases that the security returns behave neither randomly nor fuzzily. Superficially, it seems that random security returns contain objective uncertainty while fuzzy security returns reflect subjective vagueness. More deeply we will find that it is the measure which evaluates the occurrence chance of an event that differs the different kinds of security returns essentially. For example, suppose a security return is likely to be about 0.1. The occurrence chance of the security return between 0.1 and 0.2 is 30%, and the occurrence chance of the security return between 0.2 and 0.3 is 20% (see Fig. 1.1). Then what do you think the occurrence chance of the security return between 0.1 and 0.3 to be? If you think the occurrence chance will be 50%, you in fact are believing that the security return can be described by random variable; if you think the occurrence chance will be 30%, you in fact are believing that the security return should be described by fuzzy variable. However, if you think the occurrence chance should not be as big as 50%, nor should it be as small as 30%, instead, it should be a number between 30% and 50%, then you in fact are believing that the security return can be described by another kind of variable, i.e., uncertain variable.

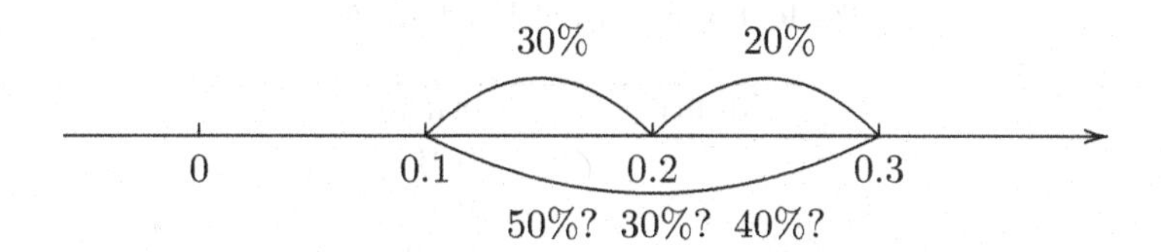

Fig. 1.1 Security return may behave neither randomly nor fuzzily.

For the definitions and the useful knowledge of *random variable*, *fuzzy variable*, *uncertain variable* and the applications of them in portfolio selection, we will introduce in later chapters.

1.2 Portfolio Return

Suppose we have n numbers of alternative securities. Let ξ_i denote the i-th security returns and x_i the investment proportions in the i-th securities, $i = 1, 2, \cdots, n$, respectively. Then a portfolio return is the sum of

$$x_1\xi_1 + x_2\xi_2 + \cdots + x_n\xi_n.$$

If the security returns ξ_i are deterministic, we can tell the returns of any portfolios exactly.

For example, suppose there are 3 securities and their returns are $\xi_1 = 0.1, \xi_2 = 0.12$, and $\xi_3 = 0.15$, respectively. If we have one portfolio A in which we invest 25% of our money in Security 1, 35% of our money in Security 2, and 40% of our money in Security 3, then the portfolio return is very exact and is calculated as

$$25\% \times 0.1 + 35\% \times 0.12 + 40\% \times 0.15 = 0.127.$$

However, as discussed in Section 1.1, the security returns ξ_i may be of random, fuzzy, or uncertain characteristic rather than deterministic. Then the portfolio return

$$x_1\xi_1 + x_2\xi_2 + \cdots + x_n\xi_n$$

is quite variable. It is hard to tell exactly what value the portfolio return will be.

For example, assume we have one portfolio A, and believe that the past returns of the portfolio can well reflect its future return. The monthly return of Portfolio A is obtained and listed in Table 1.1. Still, we cannot tell clearly what level the future return of the portfolio will be because the future return of the portfolio will vary from -0.20 to 0.20. To tell clearly the portfolio return, we need to find a deterministic number that can characterize and represent the variable portfolio return.

Table 1.1 Return of Portfolio A

Month	Returns
1	0.10
2	0.11
3	0.20
4	-0.05
5	0.16
6	0.12
7	-0.10
8	0.15
9	-0.03
10	-0.20
11	0.16
12	0.10
Expected value	0.06

Expected return is the deterministic number that was first proposed by Markowitz [66] to characterize and represent the variable portfolio return. Expected return gives the average information about the variable portfolio return. According to Markowitz, the monthly return of Portfolio A can be represented by the average value

$$\frac{0.10 + 0.11 + 0.20 - 0.05 + \cdots + 0.10}{12} = 0.06.$$

Another deterministic number that is used to represent the variable portfolio return is called the β-return. If the security returns are all regarded to be random, the β-return of the portfolio is the maximum return that one can obtain at the probability not less than the preset confidence level β. That is, the β-return of the portfolio, denoted by $\xi(\beta)$, is defined by

$$\xi(\beta) = \sup\{\bar{f} | \Pr\{x_1\xi_1 + x_2\xi_2 + \cdots + x_n\xi_n \geq \bar{f}\} \geq \beta\},$$

where $\beta \in [0, 1]$ is the preset confidence level. For portfolio A in Table 1.1, if we set the confidence level $\beta = 0.8$, then the 0.8-return of Portfolio A is -0.05 because -0.05 is the maximum return that Portfolio A can reach at the probability not less than 0.80. In fact, the β-return is the percentile in probability theory. Since it is more consistent in portfolio analysis, we use the term β-return, which will also be used in fuzzy portfolio selection and in uncertain portfolio selection with security returns behaving neither randomly nor fuzzily.

Table 1.2 Return of Portfolio B

Month	Returns
1	0.12
2	0.15
3	0.16
4	-0.12
5	0.20
6	0.10
7	0.13
8	-0.04
9	0.08
10	-0.18
11	0.10
12	-0.05
Expected value	0.054

With the deterministic numbers that represent the variable portfolio returns, we are able to compare the return levels of any two portfolios. For example, we have another portfolio B whose monthly returns are listed in Table 1.2. We believe that the past returns of portfolio B can also well reflect its future return. Then, if we use expected return as the representative value of portfolios A and B, the return of portfolio A is bigger than the return of portfolio B because the expected return of portfolio A is 0.06 while the expected return of portfolio B is 0.054. If we use the 0.8-return as the representative value of the two portfolios, the returns of portfolios A and B are regarded to be the same because the 0.8-return of portfolios A and B are both 0.05. However, the return of portfolio A is bigger than the return of portfolio B if we use the 0.90-return as the representative value of the two portfolios because the 0.9-return of portfolio A is -0.10 while the 0.9-return of portfolio B is -0.12. It is seen that using different representative values will lead to different judgement results.

1.3 What Is Risk

Though every one talks about risk of investment, the measurable definition of risk was not given until 1952. In 1952, Markowitz [66] proposed that the variance of a variable portfolio return can be regarded as the investment risk. Variance is the average squared deviation from the expected value. The measure of risk assumes that the greater deviation from the expected return, the more likely the investors cannot obtain the expected return.

For example, for the above-mentioned Portfolio A in Table 1.1, its variance is:

$$\frac{0.0016 + 0.0025 + 0.0196 + \cdots + 0.0016}{12} = 0.014.$$

That is, the risk level of Portfolio A is 0.014 (see Table 1.3).

Table 1.3 Computation of Variance

Month	Returns	Deviations from expected return	Squared deviations
1	0.10	0.04	0.0016
2	0.11	0.05	0.0025
3	0.20	0.14	0.0196
4	-0.05	-0.11	0.0121
5	0.16	0.10	0.01
6	0.12	0.06	0.0036
7	-0.10	-0.16	0.0256
8	0.15	0.09	0.0081
9	-0.03	-0.09	0.0081
10	-0.20	-0.26	0.0676
11	0.16	0.10	0.01
12	0.10	0.04	0.0016
Average	0.06	0.00	0.014

Though variance is a popular definition of risk, it just provides a comprehensive information about deviation level from the expected return. Loss is not observable. Variance does not give investors any instinct information about loss. However, in reality, people usually concern the instinct loss. Furthermore, it is a common phenomenon that when people make their risk-taking or risk-avoiding decision, they are actually weighing two factors. One is the severity level of the likely loss and the other is the occurrence chance of the loss event. For example, some people do not like taking plane because though the probability of crashing (likely loss event) is very low, the severity level (loss of life) is too great to accept. This is also true in the case of investment. When evaluating if a portfolio is safe enough, some investors would evaluate the occurrence chances of all the likely losses and the severity levels of these likely losses. Therefore, Huang [34, 36, 38] proposed another quantitative definition of risk called risk curve to describe each likely loss level and the occurrence chance of the likely loss event.

To understand risk curve, let us first understand what the likely losses are. Or say how should we describe the likely losses mathematically. Suppose we have n numbers of alternative securities. We denote ξ_i the i-th security returns and x_i the investment proportions in the i-th securities, $i = 1, 2, \cdots, n$, respectively. Assume r_f is the risk-free interest rate. Then when

$$r_f - (x_1\xi_1 + x_2\xi_2 + \cdots + x_n\xi_n) \geq 0,$$

it means that the portfolio return $x_1\xi_1+x_2\xi_2+\cdots+x_n\xi_n$ is lower than the risk-free interest rate r_f, and the difference is $r_f - (x_1\xi_1+x_2\xi_2+\cdots+x_n\xi_n)$ value. This value can certainly be understood as a loss. For example, suppose the risk-free interest rate is 0. If the portfolio return is -0.05, then the investor's loss is $0-(-0.05) = 0.05$. If the portfolio return is -0.12, then the investor's loss is $0 - (-0.12) = 0.12$. Of course, the risk-free interest rate is usually higher than zero. If the risk-free interest rate is 0.02, then even when the portfolio return is 0.01, the investor will still think he or she has experienced a loss of $0.02 - 0.01 = 0.01$. Since the portfolio return is variable, the formula

$$r_f - (x_1\xi_1 + x_2\xi_2 + \cdots + x_n\xi_n) \geq r, \forall r \geq 0$$

describes all the likely losses. Therefore, the risk curve of the portfolio return $(x_1\xi_1 + x_2\xi_2 + \cdots + x_n\xi_n)$ is defined as follows:

$$R(x_1, x_2, \cdots, x_n; r) = \pi\{r_f - (x_1\xi_1 + x_2\xi_2 + \cdots + x_n\xi_n) \geq r\}, \quad \forall r \geq 0,$$

where $R(x_1, x_2, \cdots, x_n; r)$ is the risk curve, and π the measure that gauges the occurrence chance of the loss event $\{r_f - (x_1\xi_1+x_2\xi_2+\cdots+x_n\xi_n) \geq r\}$. The general trend of the risk curve is illustrated in Fig. 1.2.

For example, if the portfolio return is believed to be random, then the risk curve of the portfolio is expressed as follows:

$$R(x_1, x_2, \cdots, x_n; r) = \Pr\{r_f - (x_1\xi_1 + x_2\xi_2 + \cdots + x_n\xi_n) \geq r\}, \quad \forall r \geq 0.$$

According to the definition of risk curve, a portfolio is safer than another portfolio if its risk curve is lower than the risk curve of another portfolio. Fig. 1.3 gives two risk curves of hypothetical portfolios B and C. Since the risk curve of portfolio B, i.e., $R_B(r)$, is below the risk curve of portfolio C, i.e., $R_C(r)$, portfolio B is safer than portfolio C.

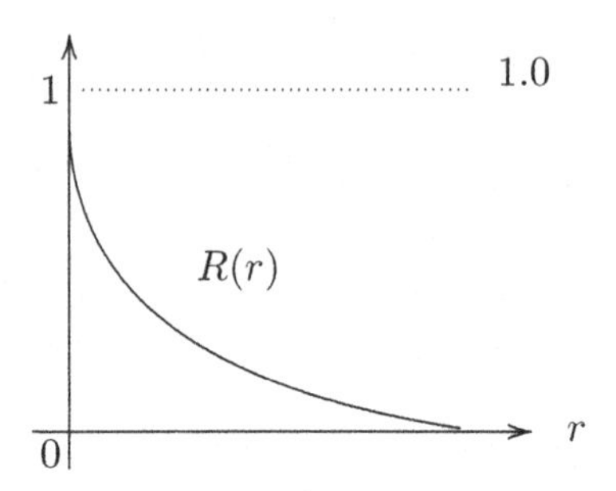

Fig. 1.2 General trend of a risk curve. The greater the r value, the less the $R(r)$ value.

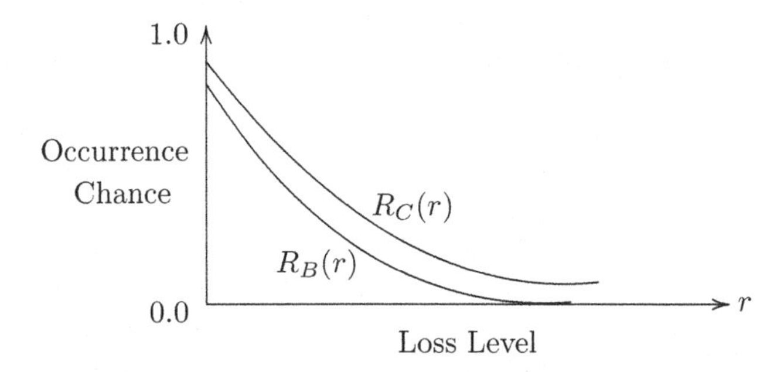

Fig. 1.3 Risk curves of two portfolios B and C.

Sometimes, the investors may only be sensitive to one preset disastrous loss level instead of all the likely loss levels. Then they will regard the occurrence chance of the specific loss level as the risk. That is, for portfolio $(x_1\xi_1+x_2\xi_2+\cdots+x_n\xi_n)$, if the investors set the disastrous loss level at r_0, then the risk now is defined as

$$\pi\{r_f - (x_1\xi_1 + x_2\xi_2 + \cdots + x_n\xi_n) \geq r_0\}$$

where r_f is the risk-free interest rate, and π the measure that gauges the occurrence chance of the loss event $\{r_f - (x_1\xi_1 + x_2\xi_2 + \cdots + x_n\xi_n) \geq r_0\}$.

For example, if the portfolio return is believed to be random, the risk of the portfolio now is

$$\Pr\{r_f - (x_1\xi_1 + x_2\xi_2 + \cdots + x_n\xi_n) \geq r_0\}.$$

For the above mentioned Portfolio A in Table 1.1, if the risk-free interest rate is 0.01, and the investors set the specific disastrous loss level at $r_0 = 0.1$, then the risk level of Portfolio A is $1/6$.

With the mathematical definitions of risk, we are able to compare the riskiness degrees of any two portfolios.

1.4 Portfolio Analysis and IRR Graph

Portfolios can be analyzed and judged to be "good" or "bad" when the deterministic representative numbers of the portfolio returns and the quantitative definitions of portfolio risks have been given. The **IRR** Graph in Fig. 1.4 gives three basic aspects for portfolio analysis. The first aspect is **I**nformation. We need to know the underlying security returns first. Are they random, fuzzy, or uncertain variables? The second aspect is what we use as the representative number of portfolio **R**eturn. Do we use expected return or β-return as the representative of the portfolio return? The third aspect is concerned with the definition of **R**isk. Which do we regard as the investment risk, risk curve, variance, or occurrence chance of a specific loss level? Different people have different choices, and different choices produce different results. However, any portfolio selection decision is made in the coordinate system of **I**nformation, **R**eturn and **R**isk. For example, the plane "**I**=Stochastic" represents that the portfolio return is believed to be random variable. The plane "**R**=Expected Return" means that the expected return is used to represent the portfolio return. And the plane "**R**=Risk Curve" implies that risk curve of the portfolio is regarded to be the investment risk. The point "(**I**,**R**,**R**)=(Stochastic, Expected Return, Risk Curve)" represents that the decision is made when portfolio return is random and the investors use expected return as the representative of the portfolio return and regard risk curve as the portfolio risk.

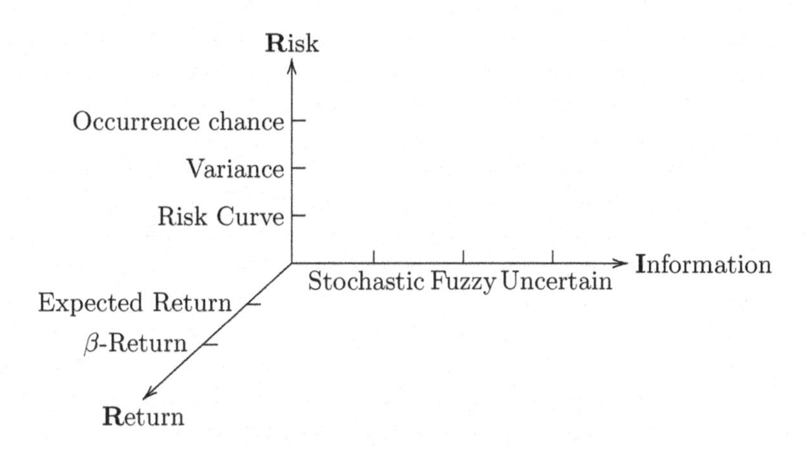

Fig. 1.4 IRR graph of portfolio analysis.

Chapter 2
Probabilistic Portfolio Selection

Probabilistic portfolio selection handles portfolio selection problem with random returns by means of probability theory. It was researched earliest and began to get rapid development since Markowitz in 1952. Before Markowitz, there were no measurable terms for risk. Mean-variance model, proposed by Markowitz [66], opened the door for mathematical analysis of portfolio selection problem. Mean-semivariance model, also proposed by Markowitz [67], served as an improvement of mean-variance selection model. As an alternative definition of risk, Roy [83] proposed probability of a preset loss level as risk, and the selection idea of minimizing the probability of a specific loss level came to be known. The nowadays popular VaR is in fact another version of Roy's risk definition. Recently, Huang [36] defined risk curve and proposed a mean-risk selection idea.

This chapter will start with review of some fundamentals of probability theory concerning probabilistic portfolio selection. Since the main concepts and results of probability theory are well-known, the credit references are not provided. Then the chapter will focus on a spectrum of portfolio selection models from different perspectives on risk and return. After that, a hybrid intelligent algorithm is documented as a general solution algorithm for the probabilistic portfolio selection model problems.

2.1 Fundamentals of Probability Theory

Random uncertainty is a basic type of uncertainty. It is usually observed in reality, especially in dice games. Probability theory originated from Pascal and Fermat's discussion on the calculations of probabilities in dice games in their famous correspondences in 1654. Later, probability theory was developed not only limited to dice games but also to all the other random phenomena. However, the rapid development of the theory did not begin until Kolmogorov presented the axiomatic foundation for probability theory in his famous book *Foundations of the Theory of Probability* in 1933. Nowadays,

X. Huang: Portfolio Analysis: From Probab. to Credibilistic, STUDFUZZ 250, pp. 11–60.
springerlink.com

probability theory has been widely applied in various fields including portfolio selection. In this subchapter, our emphasis is mainly on random variable, probability, probability distribution, probability density function, expected value, variance, semivariance, critical values and entropy.

Probability and Random Variable

Definition 2.1. *Let Ω be a nonempty set of all outcomes of a random experiment. A nonempty collection of subsets of Ω, denoted by $\mathcal{A}$, is called a σ-algebra if it has the following three properties: (i) $\Omega \in \mathcal{A}$; (ii) If $A \in \mathcal{A}$, then $A^c \in \mathcal{A}$; (iii) If $A_n \in \mathcal{A}$ and A_n is a countable sequence, then $\cup_n A_n \in \mathcal{A}$. Each element in $\mathcal{A}$ is called an event. The set function* Pr *is called a probability measure if*
(Axiom 1) (Normality) $\Pr\{\Omega\} = 1$;
(Axiom 2) (Nonnegativity) $\Pr\{A\} \geq 0$ *for any* $A \in \mathcal{A}$;
(Axiom 3) (Countable Additivity) $\Pr\left\{\bigcup_{i=1}^{\infty} A_i\right\} = \sum_{i=1}^{\infty} \Pr\{A_i\}$ *for every countable sequence of mutually disjoint events* $\{A_i\}_{i=1}^{\infty}$.

The value of $\Pr\{A\}$ indicates the probability value that the event A will occur.

Theorem 2.1. *Let Ω be a nonempty set, $\mathcal{A}$ a σ-algebra over Ω, and* Pr *a probability measure. Then we have*
(i) $\Pr\{\emptyset\} = 0$;
(ii) $0 \leq \Pr\{A\} \leq 1$ *for any* $A \in \mathcal{A}$;
(iii) Pr *is increasing, i.e.,* $\Pr\{A\} \leq \Pr\{B\}$ *whenever* $A \subset B$;
(iv) Pr *is self-dual, i.e.,* $\Pr\{A\} + \Pr\{A^c\} = 1$ *for any* $A \in \mathcal{A}$.

Definition 2.2. *Let Ω be a nonempty set, $\mathcal{A}$ a σ-algebra of subsets of Ω, and* Pr *a probability measure. Then the triplet $(\Omega, \mathcal{A}, \Pr)$ is called a probability space.*

Definition 2.3. *A random variable is a measurable function from a probability space $(\Omega, \mathcal{A}, \Pr)$ to the set of real numbers.*

Remark 2.1. Let Ω be a nonempty set, and $\mathcal{A}$ a σ-algebra over Ω. Then $(\Omega, \mathcal{A})$ is called a measurable space, and the sets in $\mathcal{A}$ are called measurable sets.

Remark 2.2. The smallest σ-algebra containing all open intervals of the set of n-dimensional real numbers $\Re^n$ is called a Borel algebra of $\Re^n$. Any element in the Borel algebra is called a Borel set. It is easy to see that any Borel sets are measurable sets.

Remark 2.3. A function f from $(\Omega, \mathcal{A})$ to the set of real numbers is said to be measurable if for any Borel set B of real numbers, we have

$$f^{-1}(B) = \{\omega \in \Omega | f(\omega) \in B\} \in \mathcal{A}.$$

Remark 2.4. Since a random variable ξ is a measurable function, for any Borel set B of real numbers, the set

$$\{\xi \in B\} = \{\omega \in \Omega | \xi(\omega) \in B\} \in \mathcal{A},$$

which means that $\{\xi \in B\}$ is an event. In practice, we usually express the event $\{\xi \in B\}$ by $\{\xi \le t\}$ or $\{\xi \ge t\}$ where t is a real number. For example, let ξ represent a random portfolio return. Then the event that the portfolio return is not less than 0.10 can be expressed by $\{\xi \ge 0.10\}$.

Example 2.1. Take $(\Omega, \mathcal{A}, \Pr)$ to be $\{\omega_1, \omega_2, \omega_3, \omega_4, \omega_5, \omega_6\}$ with $\Pr\{\omega_1\} = \Pr\{\omega_2\} = \Pr\{\omega_3\} = \Pr\{\omega_4\} = \Pr\{\omega_5\} = \Pr\{\omega_6\} = 1/6$. Then the function

$$\xi(\omega) = \begin{cases} 1, & \text{if } \omega = \omega_1 \\ 2, & \text{if } \omega = \omega_2 \\ 3, & \text{if } \omega = \omega_3 \\ 4, & \text{if } \omega = \omega_4 \\ 5, & \text{if } \omega = \omega_5 \\ 6, & \text{if } \omega = \omega_6 \end{cases}$$

is a random variable. In reality, this random variable is usually used to describe the likely results of playing dice with six facets. From the remark 2.4, it can be seen that the event $\{\omega_i\}, i = 1, 2, 3, 4, 5, 6$, can also be represented by $\{\xi = i\}, i = 1, 2, 3, 4, 5, 6$. In fact, in practice, the event that the facet with i numbers of dot will appear is usually represented by $\{\xi = i\}$. It is easy to see that the probability that the event $\{\xi = i\}$ will occur is 1/6, i.e., $\Pr\{\xi = i\} = 1/6$.

Example 2.2. Take $\Omega = [0, 1]$ and $\mathcal{A}$ the Borel-algebra, with Pr the lebesgue measure (the measure π on the Borel algebra of real numbers such that $\pi\{(a, b]\} = b - a$ for $\forall (a, b]$ is called the Lebesgue measure). Then the function $\xi : \Omega \to [a, b]$, i.e.,

$$\xi(\omega) = a + (b - a) \cdot \omega, \quad \omega \in \Omega = [0, 1]$$

is the random variable with equi-probability distribution. It can be calculated that

$$\Pr\{\xi(\omega) \le t\} = \begin{cases} 0, & \text{if } t \le a \\ \dfrac{t - a}{b - a}, & \text{if } a \le t \le b \\ 1, & \text{if } t \ge b. \end{cases}$$

As observed in the above two examples, as a practitioner, we are not interested in the specific nature of the sample space Ω nor the specific function which defines the random variable ξ. Instead, we are interested in the values of probabilities of the events that the random variable takes some real values, for example, $\Pr\{\omega \in \Omega | \xi(\omega) \le t\}$, or simply $\Pr\{\xi \le t\}$. In probabilistic portfolio selection, we are also only interested in the probabilities that the portfolio return takes certain values.

Definition 2.4. *Let ξ_1 and ξ_2 be random variables on the probability space $(\Omega, \mathcal{A}, \Pr)$. We say $\xi_1 = \xi_2$ if $\xi_1(\omega) = \xi_2(\omega)$ for almost all $\omega \in \Omega$.*

Probability Distribution and Probability Density Function

Definition 2.5. *The probability distribution Φ: $\Re \to [0, 1]$ of a random variable ξ is defined by*

$$\Phi(t) = \Pr\left\{\omega \in \Omega \;\middle|\; \xi(\omega) \le t\right\}. \tag{2.1}$$

That is, $\Phi(t)$ is the probability that the random variable ξ takes a value less than or equal to t.

Definition 2.6. *The probability density function ϕ: $\Re \to [0, +\infty)$ of a random variable ξ is a function such that*

$$\Phi(t) = \int_{-\infty}^{t} \phi(y)\mathrm{d}y \tag{2.2}$$

holds for all $t \in \Re$, where Φ is the probability distribution of the random variable ξ.

Remark 2.5. Generally speaking, two random variables $\xi_1 \ne \xi_2$ even when ξ_1 and ξ_2 have the same probability distributions. For example, take $(\Omega, \mathcal{A}, \Pr)$ to be $\{\omega_1, \omega_2\}$ with $\Pr\{\omega_1\} = \Pr\{\omega_2\} = 0.5$. Define two random variables as follows:

$$\xi_1(\omega) = \begin{cases} -1, & \text{if } \omega = \omega_1 \\ 1, & \text{if } \omega = \omega_2, \end{cases} \qquad \xi_2(\omega) = \begin{cases} 1, & \text{if } \omega = \omega_1 \\ -1, & \text{if } \omega = \omega_2. \end{cases}$$

We can find that ξ_1 and ξ_2 have the same probability distribution, i.e.,

$$\Phi(t) = \begin{cases} 0, & \text{if } t < -1 \\ 0.5, & \text{if } -1 \le t < 1 \\ 1, & \text{if } t \ge 1. \end{cases}$$

However, it is clear that $\xi_1 \ne \xi_2$ in the sense of Definition 2.4.

Remark 2.6. Since one probability distribution may be connected to several different random variables, random variable cannot be defined via probability distribution in the research of mathematical theory. However, in application, to study a random phenomenon, it is enough to begin with probability distribution or probability density function of a random variable. In probabilistic portfolio selection, we start with probability distributions or probability density functions of portfolio returns.

Expected Value

Definition 2.7. *Let ξ be a random variable. Then the expected value of ξ is defined by*

$$E[\xi] = \int_0^{+\infty} \Pr\{\xi \geq t\}\mathrm{d}t - \int_{-\infty}^{0} \Pr\{\xi \leq t\}\mathrm{d}t \tag{2.3}$$

provided that at least one of the two integrals is finite.

Theorem 2.2. *Let ξ be a random variable whose probability density function ϕ exists. If the Lebesgue integral*

$$\int_{-\infty}^{+\infty} t\phi(t)\mathrm{d}t$$

is finite, then we have

$$E[\xi] = \int_{-\infty}^{+\infty} t\phi(t)\mathrm{d}t. \tag{2.4}$$

Expected value measures *central tendency.* It tells us where the center of the distribution of a random variable is located.

Theorem 2.3. *Let ξ and η be random variables with finite expected values. For any numbers a and b, we have $E[a\xi + b\eta] = aE[\xi] + bE[\eta]$.*

That is, the expected value operator has the linearity property.

Variance and Semivariance

Variance is the average of the squared deviations from the expected value.

Definition 2.8. *Let ξ be a random variable with finite expected value e. Then the variance of ξ is defined by $V[\xi] = E[(\xi - e)^2]$, and $\sqrt{V[\xi]}$ is called standard deviation.*

Definition 2.9. *Let ξ be a random variable with finite expected value e. Then the semivariance of ξ is defined by $SV[\xi] = E[[(\xi - e)^-]^2]$ where*

$$(\xi - e)^- = \begin{cases} \xi - e, & \text{if } \xi \leq e \\ 0, & \text{if } \xi > e. \end{cases}$$

The semivariance of a random variable measures only the negative deviations of the distribution from the expected value. A small value of semivariance indicates that the mean squared negative deviations of the random variable from its expected value is small; and a large value of semivariance indicates that the mean squared negative deviations of the random variable from its expected value is large.

Three Special Types of Random Variable

Uniform Random Variable: A random variable ξ is called a uniform random variable if its probability density function is defined by

$$\phi(t) = \begin{cases} \dfrac{1}{b-a}, & \text{if } a \leq t \leq b \\ 0, & \text{otherwise,} \end{cases} \tag{2.5}$$

where a and b are given real numbers with $a < b$. We denote the variable by $\xi \sim \mathcal{U}(a, b)$.

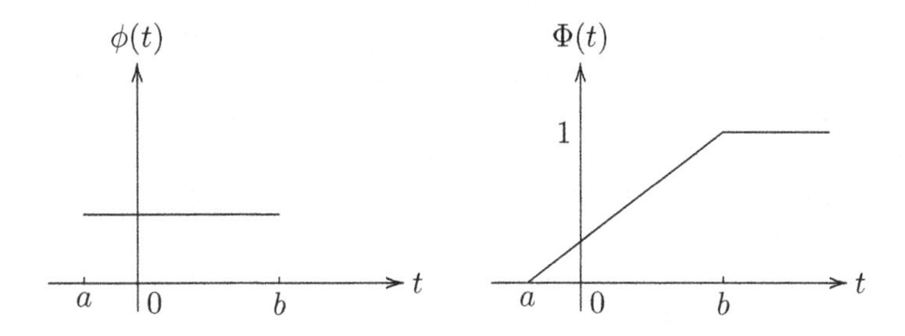

Fig. 2.1 Density function and probability distribution of $\xi \sim \mathcal{U}(a, b)$.

Example 2.3. Let ξ be a uniform random variable on the interval $[a, b]$. Then its expected value is $E[\xi] = (a + b)/2$ and variance is $V = (b - a)^2/12$.

From Equation (2.2), it is easy to see that for any two sets $A_1 = (a_1, b_1) \subseteq [a, b]$ and $A_2 = (a_2, b_2) \subseteq [a, b]$, their probabilities are the same if $b_1 - a_1 = b_2 - a_2$. That is, $\Pr\{A_1\} = \Pr\{A_2\}$ if $b_1 - a_1 = b_2 - a_2$.

In probabilistic portfolio selection, if we have no idea at all about the return of an initial public offering, but can only guess the lowest and highest values that the security return can reach, according to Laplace principle, the probabilities that the security return lies in the intervals with the same length should take the same values. Thus, we can use the uniform random variable to describe this security return.

Theorem 2.4. *Let ξ be a uniform random variable $\xi \sim \mathcal{U}(a, b)$ and λ a real number. Then we have*

$$\lambda \cdot \xi \sim \begin{cases} \mathcal{U}(\lambda a, \lambda b), & \text{if } \lambda > 0 \\ \mathcal{U}(\lambda b, \lambda a), & \text{if } \lambda < 0. \end{cases} \tag{2.6}$$

That is, the product of a uniform random variable and a scalar number is also a uniform random variable.

Theorem 2.5. *Let ξ_1 and ξ_2 be two uniform random variables $\xi_1 \sim \mathcal{U}(a_1, b_1)$ and $\xi_2 \sim \mathcal{U}(a_2, b_2)$. Then we have*

$$\xi_1 + \xi_2 \sim \mathcal{U}(a_1 + a_2, b_1 + b_2). \tag{2.7}$$

That is, the sum of two uniform random variables is also a uniform random variable. Thus, a weighted sum of uniform random variables is also a uniform random variable.

Example 2.4. Assume ξ_1 and ξ_2 are uniform random variables $\xi_1 \sim \mathcal{U}(a_1, b_1)$ and $\xi_2 \sim \mathcal{U}(a_2, b_2)$ and λ_1 and λ_2 real numbers. Then we have

$$\lambda_1 \cdot \xi_1 + \lambda_2 \cdot \xi_2 \sim \begin{cases} \mathcal{U}(\lambda_1 a_1 + \lambda_2 a_2, \lambda_1 b_1 + \lambda_2 b_2), & \text{if } \lambda_1 > 0,\ \lambda_2 > 0 \\ \mathcal{U}(\lambda_1 b_1 + \lambda_2 b_2, \lambda_1 a_1 + \lambda_2 a_2), & \text{if } \lambda_1 < 0,\ \lambda_2 < 0 \\ \mathcal{U}(\lambda_1 b_1 + \lambda_2 a_2, \lambda_1 a_1 + \lambda_2 b_2), & \text{if } \lambda_1 < 0,\ \lambda_2 > 0 \\ \mathcal{U}(\lambda_1 a_1 + \lambda_2 b_2, \lambda_1 b_1 + \lambda_2 a_2), & \text{if } \lambda_1 > 0,\ \lambda_2 < 0. \end{cases}$$

Normal Random Variable: A random variable ξ is called a normal random variable if its probability density function is defined as:

$$\phi(t) = \frac{1}{\sigma\sqrt{2\pi}} \exp\left[-\frac{(t-\mu)^2}{2\sigma^2}\right], \quad \sigma > 0, \quad t \in \Re, \tag{2.8}$$

where μ is a real number. We denote the variable by $\xi \sim \mathcal{N}(\mu, \sigma^2)$. The probability density function and probability distribution function of a normal random variable are drawn in Fig. 2.2.

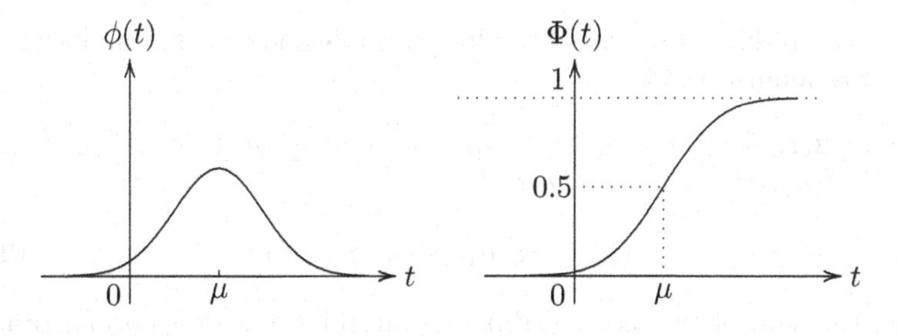

Fig. 2.2 Density function and probability distribution of $\xi \sim \mathcal{N}(\mu, \sigma^2)$.

Example 2.5. Let ξ be a normal random variable $\xi \sim \mathcal{N}(\mu, \sigma^2)$. Then its expected value is $E[\xi] = \mu$ and variance is $V[\xi] = \sigma^2$.

The probability density functions of normal random variables with different values of expected value μ and standard deviation σ are shown in Figs. 2.3 and 2.4.

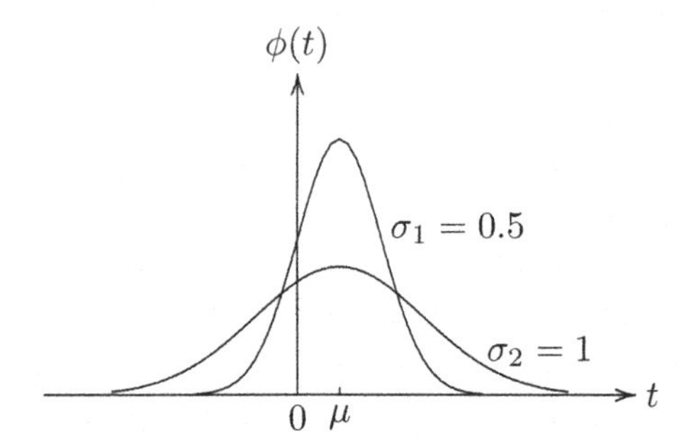

Fig. 2.3 Density functions of two normal random variables with the same μ but different σ's.

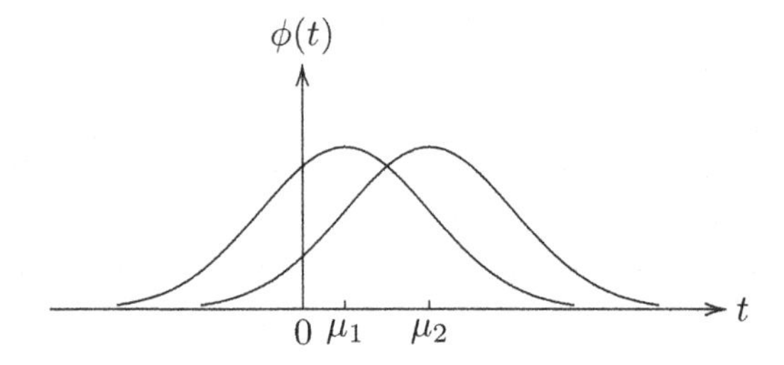

Fig. 2.4 Density functions of two normal random variables with the same σ but different μ's.

Theorem 2.6. *Let ξ be a normal random variable $\mathcal{N}(\mu, \sigma^2)$ and λ a real number. Then we have*

$$\lambda \cdot \xi \sim \mathcal{N}(\lambda\mu, \lambda^2\sigma^2) \tag{2.9}$$

That is, the product of a normal random variable and a scalar number is also a normal random variable.

Theorem 2.7. *Let ξ_1 and ξ_2 be two normal random variables $\xi_1 \sim \mathcal{N}(\mu_1, \sigma_1^2)$ and $\xi_2 \sim \mathcal{N}(\mu_2, \sigma_2^2)$. Then we have*

$$\xi_1 + \xi_2 \sim \mathcal{N}(\mu_1 + \mu_2, \sigma_1^2 + \sigma_2^2). \tag{2.10}$$

That is, the sum of two normal random variables is also a normal random variable. Thus a weighted sum of normal random variables is also a normal random variable.

Example 2.6. Assume ξ_1 and ξ_2 are normal random variables $\xi_1 \sim \mathcal{N}(\mu_1, \sigma_1^2)$ and $\xi_2 \sim \mathcal{N}(\mu_2, \sigma_2^2)$ and λ_1 and λ_2 real numbers. Then we have

$$\lambda_1 \cdot \xi_1 + \lambda_2 \cdot \xi_2 \sim \mathcal{N}(\lambda_1\mu_1 + \lambda_2\mu_2, \lambda_1^2\sigma_1^2 + \lambda_2^2\sigma_2^2).$$

Probabilistic portfolio theory, for the most part, assumes that security returns are normally distributed, or even if individual security returns are not exactly normal, the large portfolio return will resemble a normal distribution quite closely. Some observations supported the assumption. For example, Table 2.1 summarizes the research results of one-year investments in randomly selected portfolios from NYSE stocks by Fisher and Lorie [18]. The table shows that the returns of 32-stock portfolio began to resemble the normal distribution, and the return distribution of 128-stock portfolio is virtually identical to the hypothetical normally distributed portfolio.

Table 2.1 Frequency Distribution of Rates of Return from a One-Year Investment in Randomly Selected Portfolios from NYSE-Listed Stocks

Statistic	N=1 Observed	N=1 Normal	N=8 Observed	N=8 Normal
5th percentile	-14.4	-39.2	8.1	4.6
20th percentile	-0.5	6.3	16.3	16.1
50th percentile	19.6	28.2	26.4	28.2
70th percentile	38.7	49.7	33.8	35.7
95th percentile	96.3	95.6	54.3	51.8
Minimum	-71.1	NA	-12.4	NA
Maximum	442.6	NA	136.7	NA
Mean	28.2	28.2	28.2	28.2
Standard Deviation	41.0	41.0	14.4	14.4
Skewness	255.4	0.0	88.7	0.0
Sample size	1,227	–	131,072	–
Statistic	**N=32 Observed**	**N=32 Normal**	**N=128 Observed**	**N=128 Normal**
5th percentile	17.4	16.7	22.7	22.6
20th percentile	22.2	22.3	25.3	25.3
50th percentile	27.8	28.2	28.1	28.2
70th percentile	31.6	32.9	30.0	30.0
95th percentile	40.9	39.9	34.1	33.8
Minimum	6.5	NA	16.4	NA
Maximum	73.7	NA	43.1	NA
Mean	28.2	28.2	28.2	28.2
Standard Deviation	7.1	7.1	3.4	3.4
Skewness	44.5	0.0	17.7	0.0
Sample size	32,768	–	16,384	–

Lognormal Random Variable: A random variable ξ is called a lognormal random variable if $\ln(\xi)$ is a normal random variable with expected value μ and standard deviation σ. Equivalently,

$$\xi = \exp(\eta)$$

where η is a normal random variable with expected value μ and standard deviation σ. We denote the lognormal random variable by $\xi \sim \ln\mathcal{N}(\mu, \sigma^2)$. Its probability density function is given by:

$$\phi(t) = \frac{1}{t\sigma\sqrt{2\pi}} \exp\left[-\frac{(\ln t - \mu)^2}{2\sigma^2}\right], \quad \sigma > 0, \quad t > 0. \tag{2.11}$$

Note that the random variable $\ln\xi$ follows a normal distribution, but the random variable ξ follows a lognormal distribution. The difference of the normal probability density function from the lognormal probability density function is not only the replacement of t by $\ln(t)$ but also an additional t factor in $t(2\pi)^{1/2}\sigma$. The shape of the probability density functions of the lognormal random variables $\xi \sim \ln\mathcal{N}(\mu, \sigma^2)$ with different values of μ and σ are given in Figs. 2.5 and 2.6.

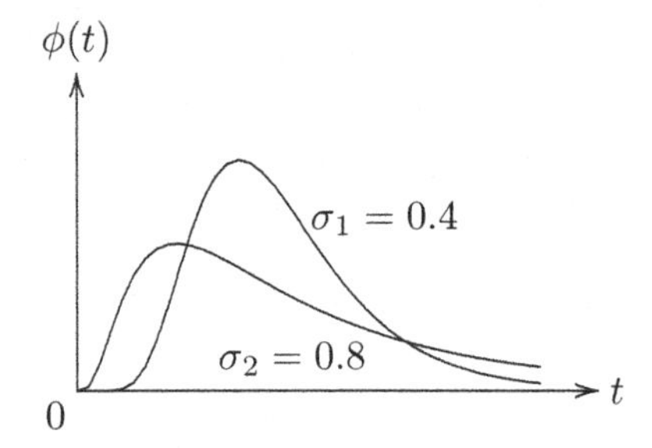

Fig. 2.5 Density functions of two lognormal random variables with the same μ but different σ's.

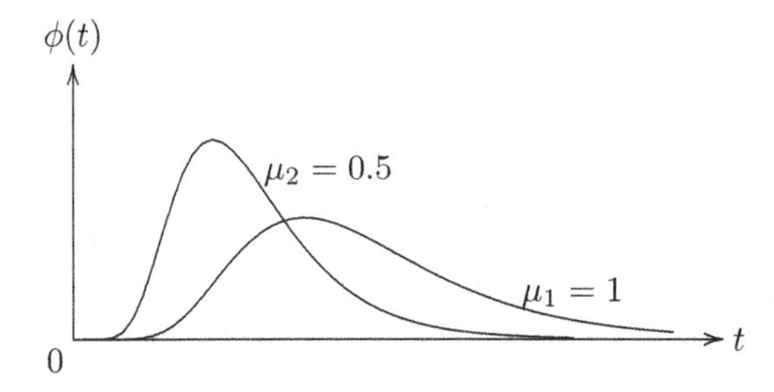

Fig. 2.6 Density functions of two lognormal random variables with the same σ but different μ's.

Since the security prices cannot be negative, there are theoretical objections to the assumption that individual security returns are normally distributed. Some people argue that the normal distribution cannot be truly representative of the security rates of return because it allows for any outcome, including the whole range of negative prices. Rates of return lower than -100% are theoretically impossible because they imply the possibility of negative security prices. To overcome the shortcoming of normal distribution assumption, an alternative assumption is that the continuously compounded annual rate of return is normally distributed. Let η represent the continuously compounded annual rate of return and ξ the effective annual rate of return. Then $\xi = \exp(\eta) - 1$. Since the smallest likely value for ξ is -1, this assumption rules out the likely outcome of negative security price. We can see that the distribution of ξ is lognormal. The famous Black-Scholes Option Pricing Formula is derived based on the lognormal distribution assumption.

However, when the holding period t is short, the approximation of the lognormal random variable $\xi(t) = \exp(\eta t) - 1$ by the normal random variable ηt is quite accurate. In other words, when the holding period is short, the normal distribution provides a good approximation to the lognormal distribution. Thus, for example, if the variance of the compounded annual rate of return is 0.12, then for all practical purposes, the variance of the one month return rate is

$$\sigma^2(\text{monthly}) = \frac{0.12}{12} = 0.01.$$

Example 2.7. Let ξ be a lognormal random variable $\xi \sim \ln\mathcal{N}(\mu, \sigma^2)$. Then its expected value is $E[\xi] = e^{\mu+\sigma^2/2}$ and variance is $V[\xi] = (e^{\sigma^2} - 1) \cdot e^{2\mu+\sigma^2}$.

Example 2.8. Let $\xi_i \sim \ln\mathcal{N}(\mu_i, \sigma_i^2), i = 1, 2, \cdots, n$, respectively, be independent lognormal random variables, and $\eta = \prod_{i=1}^{n} \xi_i$. Then η is a lognormal random variable as well:

$$\eta \sim \ln\mathcal{N}(\sum_{i=1}^{n} \mu_i, \sum_{i=1}^{n} \sigma_i^2).$$

β-Value

Definition 2.10. *Let ξ be a random variable, and $\beta \in (0, 1]$. Then*

$$\xi_{\sup}(\beta) = \sup\left\{r \mid \Pr\{\xi \geq r\} \geq \beta\right\} \tag{2.12}$$

is called the β-value of ξ.

If ξ represents a portfolio return, then the β-value of ξ means the maximum return the investors can get at least β of time (see Figs. 2.7 and 2.8). It is also called percentile in probability theory. In portfolio selection, we call it β-return because it is more meaningful for investors and more consistent for portfolio analysis.

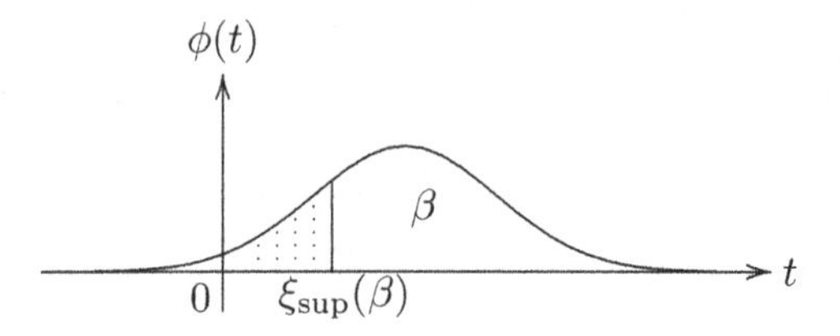

Fig. 2.7 Density function and β-value.

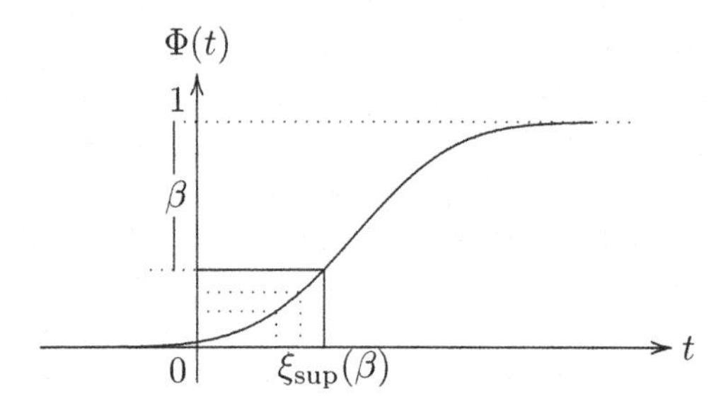

Fig. 2.8 Probability distribution and β-value.

Example 2.9. Let ξ be a normal random variable $\mathcal{N}(0,1)$. Then $\xi_{\rm sup}(0.80) = -0.84$ because $\Pr\{\xi \geq -0.84\} = 0.80$.

Theorem 2.8. *Let $\xi_{\rm sup}(\beta)$ be the β- value of the random variable ξ. Then $\xi_{\rm sup}(\beta)$ is a decreasing function of β.*

Entropy

Entropy was first defined by Shannon [85] as a measure of uncertainty. It measures the degree of difficulty of predicting the specified value that a random variable will take.

Entropy of Discrete Random Variables

Definition 2.11. *A random variable is said to be discrete if there exists a countable sequence $\{t_1, t_2, \cdots\}$ such that*

$$\Pr\{\xi \neq t_1, \xi \neq t_2, \cdots\} = 0.$$

Definition 2.12. *Let ξ be a discrete random variable taking values t_i with probability $p_i, i = 1, 2, 3, \cdots$, respectively. Then its entropy is defined by*

$$H[\xi] = -\sum_{i=1}^{\infty} p_i \ln p_i. \tag{2.13}$$

Please note that the entropy relies only on the number of values and their probabilities and does not rely on the actual values that the random variable takes.

2.2 Mean-Risk Model

Consider one case. In the case, people may lose 1 dollar at probability 99% and lose 1,000,000 dollars at probability 1%. Will they think the event is risky? Some people may say no because they can tolerate 1,000,000 dollars loss at a very low occurrence probability value of 1% while other people may say yes because to them the loss of 1,000,000 dollars is too great to tolerate even at the low occurrence probability value of 1%. The phenomenon implies that when the investors make their risk-taking or risk-avoiding decision, they are actually weighing two factors. One is severity level of a loss and the other the occurrence chance of the loss. To reflect the attitude towards risk, Huang [36] defined risk curve and proposed the mean-risk model.

2.2.1 Risk Curve

Usually, when an investment return is -0.1, people will instinctively feel that they suffer a loss of 0.1. This in fact implies that people set their breakeven point at 0 and they experience a difference, i.e., $0-(-0.1)$, of the investment return from the point. In portfolio investment, since the portfolio return is variable and may be -0.05, -0.11, $\cdots$, etc., people' loss may be 0.05, 0.11, $\cdots$, etc. When the portfolio return is 0.1, people will feel they gain and now the difference, i.e., $0-0.1$, of the portfolio return from the breakeven point is a negative number. Thus, it is clear that if ξ represents the variable portfolio return, then $0-\xi$ describes all the likely losses when $0-\xi \geq 0$. Of course, the investors can set their breakeven point higher than 0. In financial investment, people have a choice to invest their money in risk-free asset and gain a return rate as high as the risk-free interest rate with certainty. Thus, the risk-free interest rate, denoted by r_f, is chosen as the breakeven point in portfolio analysis in our book. Then, if the risk-free interest rate is 0.015, the investors will still suffer a loss of $0.015-0.01=0.005$ even when the portfolio return is 0.01. Taking into account all the likely losses of a portfolio and the corresponding occurrence chances of these losses, we define the risk curve as follows:

Definition 2.13. *(Huang [36]) Let ξ denote a random return of a portfolio, and r_f the risk-free interest rate. Then the curve*

$$R(r)=\Pr\{r_f-\xi \geq r\}, \ \forall r \geq 0 \tag{2.14}$$

is called the risk curve of the portfolio, and r the loss severity indicator.

In the definition, the set $\{r_f-\xi \geq r\}$ describes the event that the loss is equal to or greater than the value r. The greater the value r, the severer the loss $r_f-\xi$ is. The curve $R(r)$ is the probability of the loss equal to or greater than the value r. Note that r is a nonnegative value rather than a fixed number. Then $R(r) = \Pr\{r_f - \xi \geq r\}$ is a curve which gives all the probabilities of all the likely losses.

Example 2.10. Let ξ be the random return of a portfolio with normal distribution, i.e., $\xi \sim \mathcal{N}(\mu, \sigma^2)$. Then the risk curve of ξ is expressed as follows:

$$R(r) = \Pr\{(r_f - \xi) \geq r\} = \frac{1}{\sigma\sqrt{2\pi}} \int_{-\infty}^{r_f - r} \exp\left[-\frac{(t-\mu)^2}{2\sigma^2}\right] \mathrm{d}t, \quad r \geq 0.$$

The curve is shown in Fig. 2.9.

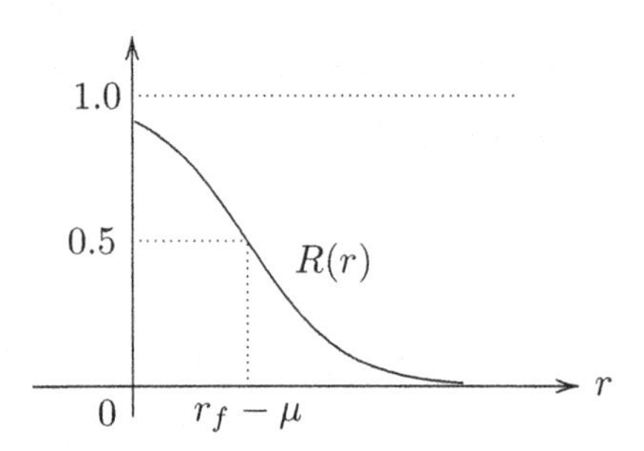

Fig. 2.9 Risk curve of a portfolio with normally distributed return.

Example 2.11. Let ξ be the random return of a portfolio with uniform distribution, i.e., $\xi \sim \mathcal{U}(a, b)$. Then the risk curve of ξ is as follows:

$$R(r) = \Pr\{(r_f - \xi) \geq r\} = \begin{cases} 1, & \text{if } r < r_f - b \\ \dfrac{r_f - a - r}{b - a}, & \text{if } r_f - b \leq r \leq r_f - a \\ 0, & \text{if } r > r_f - a. \end{cases}$$

The curve is shown in Fig. 2.10.

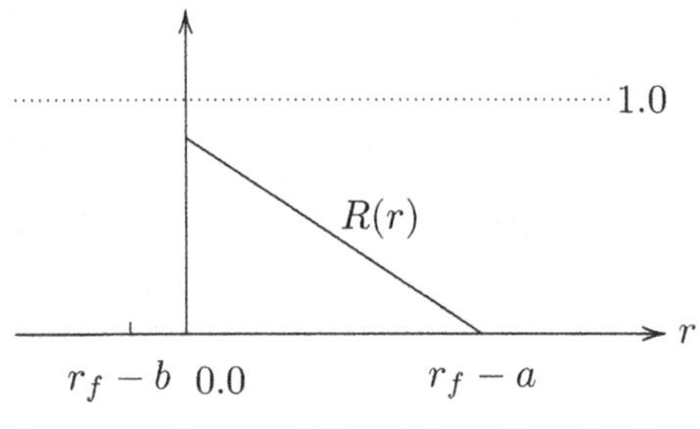

Fig. 2.10 Risk curve of a portfolio with uniformly distributed return.

Example 2.12. Risk curves of Securities Baosteel, CITIC, and their combination

According to the data published by Shanghai Stock Market, we calculate the monthly returns of Securities Baosteel (600019) and CITIC (600030) and list the returns in Table 2.2. The monthly return is calculated by

$$\frac{\text{(Closing price, this month)-(Closing price, last month)}}{\text{(Closing price, last month)}}$$

where the closing price of each month is the adjusted price having considered stock split and dividend during the month. For example, the adjusted closing price of Baosteel in January 2006 is RMB 3.71, and the adjusted closing price of Baosteel in December 2005 is RMB 3.74. Then the return rate of Baosteel in January 2006 is (3.71-3.74)/3.74=-0.01.

Table 2.2 Returns and Losses of Baosteel and CITIC

Month	Bao. Return	Bao. Loss	CITIC Return	CITIC Loss
1/2006	-0.01	0.013	0.35	-
2/2006	0.06	-	-0.09	0.093
3/2006	-0.03	0.033	0.19	-
4/2006	0.00	0.003	0.55	-
5/2006	0.13	-	0.29	-
6/2006	-0.02	0.023	0.04	-
7/2006	-0.07	0.073	-0.15	0.153
8/2006	0.02	-	0.04	-
9/2006	0.00	0.003	0.08	-
10/2006	0.18	-	-0.02	0.023
11/2006	0.40	-	0.27	-
12/2006	0.27	-	0.46	-
1/2007	0.13	-	0.28	-
2/2007	-0.03	0.033	0.06	-
3/2007	0.05	-	0.16	-
4/2007	0.13	-	0.37	-
5/2007	0.10	-	-0.08	0.083
6/2007	-0.08	0.083	-0.02	0.023
7/2007	0.23	-	0.24	-
8/2007	0.37	-	0.36	-
9/2007	-0.02	0.023	0.09	-
10/2007	0.02	-	0.10	-
11/2007	-0.22	0.223	-0.21	0.213
12/2007	0.20	-	0.06	-

Table 2.3 Loss Frequencies of Baosteel and CITIC

Loss Indicator r	Baosteel (Loss $\geq r$) Relative Frequency	CITIC (Loss $\geq r$) Relative Frequency
0.00	10/24	6/24
0.01	8/24	6/24
0.02	7/24	6/24
0.03	5/24	4/24
0.04	3/24	4/24
0.05	3/24	4/24
0.06	3/24	4/24
0.07	3/24	4/24
0.08	2/24	4/24
0.09	1/24	3/24
0.10	1/24	2/24
0.11	1/24	2/24
0.12	1/24	2/24
0.13	1/24	2/24
0.14	1/24	2/24
0.15	1/24	2/24
0.16	1/24	1/24
0.17	1/24	1/24
0.18	1/24	1/24
0.19	1/24	1/24
0.20	1/24	1/24
0.21	1/24	1/24
0.22	1/24	0

Table 2.4 Loss Frequencies of Portfolio BC

Loss Level (r)	Relative Frq. (Loss $\geq r$)	Loss Level (r)	Relative Frq. (Loss $\geq r$)
0.00	4/24	0.11	1/24
0.01	4/24	0.12	1/24
0.02	3/24	0.13	1/24
0.03	3/24	0.14	1/24
0.04	3/24	0.15	1/24
0.05	3/24	0.16	1/24
0.06	2/24	0.17	1/24
0.07	2/24	0.18	1/24
0.08	2/24	0.19	1/24
0.09	2/24	0.20	1/24
0.10	2/24	0.21	1/24

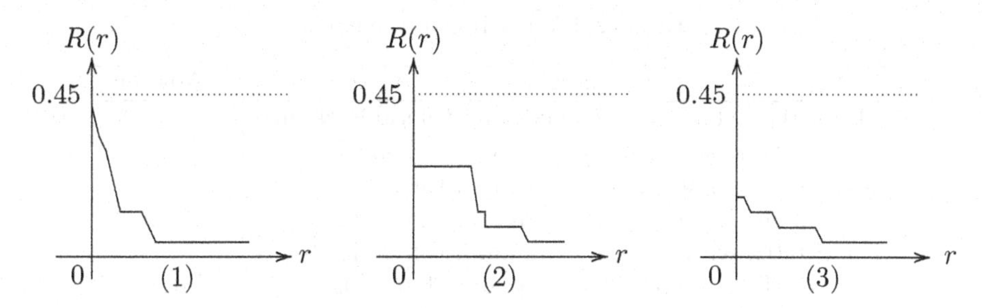

Fig. 2.11 Risk curves of securities Baosteel (the 1st one), CITIC (the 2nd one), and the portfolio BC (the 3rd one).

Suppose the monthly risk-free interest rate is $r_f = 0.003$. Then the loss of Baosteel in January 2006 is 0.003-(-0.01)=0.013. The losses of Baosteel and CITIC from January 2006 to December 2007 are given in Table 2.2. According to the data in Table 2.2, given different values of r, the relative frequencies of the loss equal to or greater than r are obtained in Table 2.3. The risk curves of securities Baosteel (the first one) and CITIC (the second one) are drawn in Fig. 2.11.

If we invest half the money in security Baosteel and half in security CITIC, then we have a portfolio. Let us call it Portfolio BC. The relative frequencies of the loss of the portfolio equal to or greater than the value r are given in Table 2.4. The risk curve of Portfolio BC is drawn in Fig. 2.11 (the third one). It is seen that the risk curve of the portfolio tends to be lower than the risk curves of the individual securities.

2.2.2 Confidence Curve and Safe Portfolio

Since all investors know that they may lose as well as gain in investment, they will have a maximum tolerance towards occurrence chance of each likely loss level. We call it confidence curve $\alpha(r)$ that gives the investors' maximal tolerance towards the occurrence chance of each likely loss level. Different investors have different tolerable occurrence chance levels even towards the same loss level. However, given any loss level r, an investor should be able to give his or her maximal tolerable chance of the loss equal to or greater than r by answering "what-if" questions. For example, if r=0, what is your maximum tolerable chance level of the loss equal to or greater than 0? If $r = 0.01$, what is your maximum tolerable chance level of the loss equal to or greater than 0.01? If $r = 0.02$, what is your maximum tolerable chance level of the loss equal to or greater than 0.02? $\cdots$ By answering all the "what-if" questions, the investor is able to give his or her maximum tolerable chances of all the likely losses (see What-If Questionnaire in Table 2.5). In probabilistic portfolio selection, occurrence chance of a random event is measured by probability measure. Though different investors' confidence curves may be in different

Table 2.5 What-If Questionnaire

Question	Answer
If $r = 0.0$, what is your maximum tolerable chance of the loss equal to or greater than 0?	1
If $r = 0.01$, what is your maximum tolerable chance of the loss equal to or greater than 0.01?	0.9
If $r = 0.02$, what is your maximum tolerable chance of the loss equal to or greater than 0.02?	0.9
If $r = 0.03$, what is your maximum tolerable chance of the loss equal to or greater than 0.03?	0.9
If $r = 0.04$, what is your maximum tolerable chance of the loss equal to or greater than 0.04?	0.8
If $r = 0.05$, what is your maximum tolerable chance of the loss equal to or greater than 0.05?	0.8
...	...
If $r = 0.98$, what is your maximum tolerable chance of the loss equal to or greater than 0.04?	0.01
If $r = 0.99$, what is your maximum tolerable chance of the loss equal to or greater than 0.05?	0.01
If $r = 1.0$, what is your maximum tolerable chance of the loss equal to or greater than 0.7?	0.0

shapes, the general trend of the curves is the same. That is, when r is low, investors can tolerate a comparatively high occurrence probability of the loss equal to or greater than r; however, when r is high, investors can tolerate only a low occurrence probability of the loss equal to or greater than this r value. Three examples of confidence curve are given below.

Example 2.13. A confidence curve can be expressed by a linear line as follows:

$$\alpha(r) = a - b \cdot r$$

where a and b are positive real numbers. Fig. 2.12 gives the confidence curve of $\alpha(r) = 0.6 - 1.2r$.

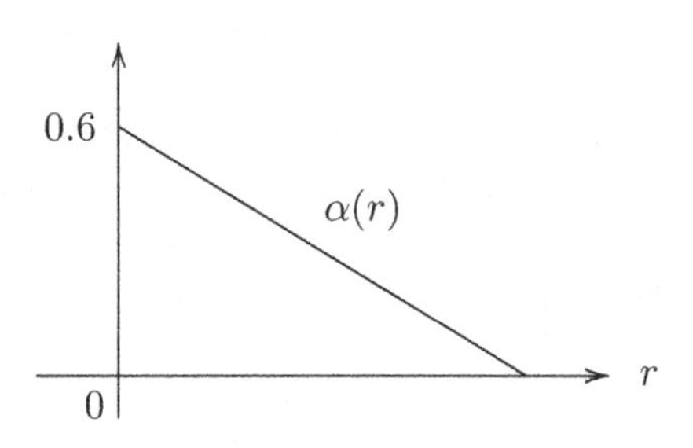

Fig. 2.12 Confidence curve of Example 2.13.

Example 2.14. A confidence curve can be expressed by a broken line as follows:

$$\alpha(r) = \begin{cases} -b_1 \cdot r + a_1, & 0 \le r \le r_1, \\ -b_2 \cdot r + a_2, & r_1 \le r \le r_2, \\ \cdots \\ -b_n \cdot r + a_n, & r \ge r_{n-1} \end{cases}$$

where $-b_i \cdot r + a_i = -b_{i+1} \cdot r + a_{i+1}$, for $r = r_i, i = 1, 2, \cdots, n-1$. Fig. 2.13 gives the confidence curve of

$$\alpha(r) = \begin{cases} -2.25r + 0.6, & 0 \le r \le 0.2, \\ -0.5r + 0.25, & 0.2 \le r \le 0.4, \\ 0.05, & r \ge 0.4. \end{cases}$$

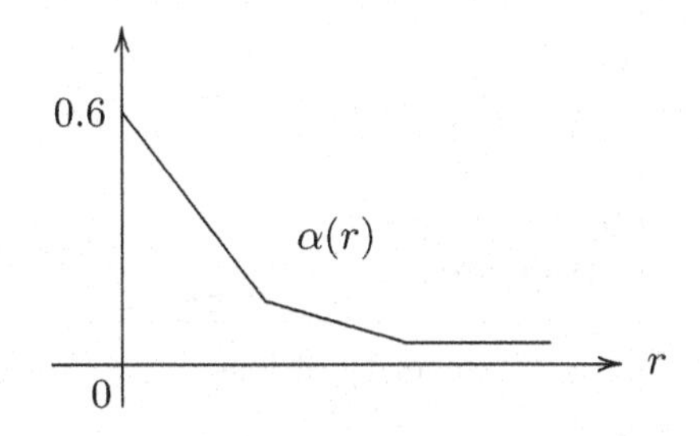

Fig. 2.13 Confidence curve of Example 2.14.

Example 2.15. Confidence curve can be expressed by a power function as follows:

$$\alpha(r) = \frac{a}{(r+1)^k}, \quad r \ge 0$$

where a is a real number and k a positive integer number. Fig. 2.14 gives the confidence curve of $\alpha(r) = \dfrac{0.6}{(r+1)^4}$, $r \ge 0$.

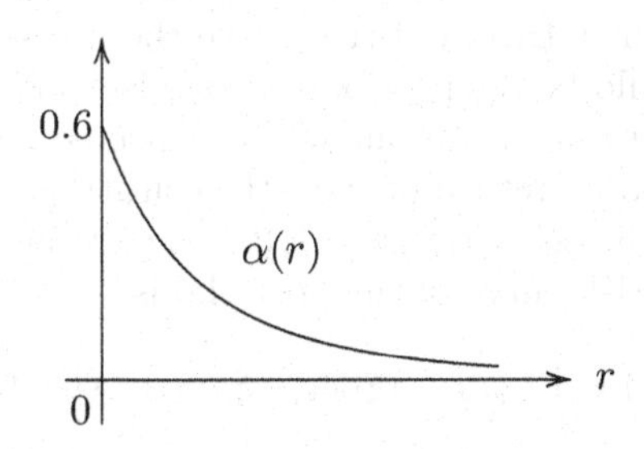

Fig. 2.14 Confidence curve of Example 2.15.

With risk curve and confidence curve, we are now able to judge if a portfolio is safe or not. It is obvious that the area below the confidence curve $\alpha(r)$ is the low risk area, and the area above the confidence curve $\alpha(r)$ the high risk area which the investor should try to avoid getting in. Thus, a portfolio is safe if its risk curve is totally below the confidence curve; a portfolio is risky if any part of its risk curve is above the confidence curve (see Fig. 2.15). The ranking criterion for riskiness of portfolios can be expressed mathematically as follows:

Let ξ be the random return of a portfolio, and $\alpha(r)$ the investor's confidence curve. We say the portfolio is safe if

$$R(r) = \Pr\{(r_f - \xi) \geq r\} \leq \alpha(r), \quad \forall r \geq 0,$$

where r_f is the risk-free interest rate.

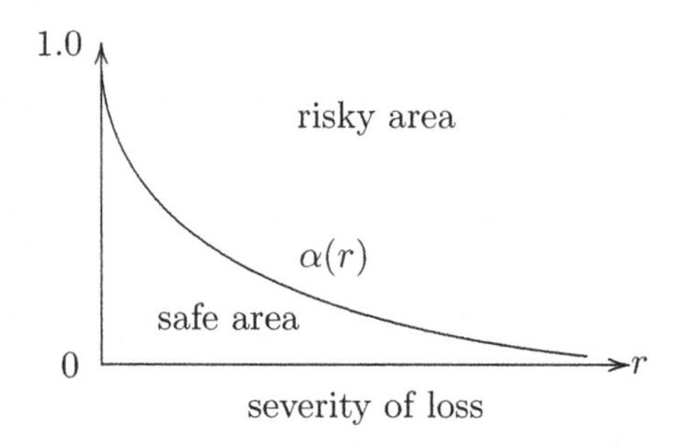

Fig. 2.15 Confidence curve and investment risk. A portfolio is safe if its risk curve is totally below the confidence curve; a portfolio is risky if any part of its risk curve is above the confidence curve.

2.2.3 Mean-Risk Model

When investing, the investors will usually first ask that the investment be safe enough. Then they will try to pursue the maximum return. The idea of mean-risk model is to regard expected return of a portfolio as the representative of investment return and risk curve the investment risk. Then the optimal portfolio should be the portfolio whose risk curve is totally below the confidence curve and in the meantime whose expected return is the maximal. Let ξ_i denote the random return of the i-th securities and x_i the investment proportions in the i-the securities, $i = 1, 2, \cdots, n$, respectively. According to Equation (2.14), the risk curve of the portfolio is

$$R(x_1, x_2, \cdots, x_n; r) = \Pr\{r_f - (\xi_1 x_1 + \xi_2 x_2 + \cdots + \xi_n x_n) \geq r\}. \quad (2.15)$$

Suppose the investors' confidence curve is $\alpha(r)$. The mean-risk selection idea is expressed mathematically as follows:

$$\begin{cases} \max E[\xi_1 x_1 + \xi_2 x_2 + \cdots + \xi_n x_n] \\ \text{subject to:} \\ \qquad R(x_1, x_2, \cdots, x_n; r) \leq \alpha(r), \ \forall r \geq 0 \\ \qquad x_1 + x_2 + \cdots + x_n = 1 \\ \qquad x_i \geq 0, \quad i = 1, 2, \cdots, n \end{cases} \tag{2.16}$$

where E is the expected value operator and $R(x_1, x_2, \cdots, x_n; r)$ the risk curve defined by formula (2.15). The constraint $R(x_1, x_2, \cdots, x_n; r) \leq \alpha(r)$, $\forall r \geq 0$ means that the probability values of all the likely losses should all be lower than the investors' tolerance levels. This constraint ensures that the optimal portfolio will be selected only among the safe portfolios. The constraint $x_i \geq 0$ implies that short sales are not allowed in the investment.

2.2.4 Application Example

To illustrate how to use the mean-risk model, let us select the portfolio from six alternative securities listed in Shanghai Stock Market. The securities are Hundsun Electronics with security code 600570, Tianjin Quanye with security code 600821, Wanwei Updated High-Tech with security code 600063, Sany Heavy Industry with security code 600031, Baosteel with security code 600019, and Tianchuang Property with security code 600791. Based on their adjusted closing prices having considered stock split and cash dividend each month from December 2005 to December 2007, monthly returns of the six securities from January 2006 to December 2007 are obtained and shown in Table 2.6. Suppose the investors give their confidence curve as follows:

$$\alpha(r) = \begin{cases} -1.25r + 0.25, & \text{when} \quad 0 \leq r \leq 0.12 \\ -0.5r + 0.16, & \text{when} \quad 0.12 \leq r \leq 0.3 \\ 0.01, & \text{when} \quad r \geq 0.3. \end{cases}$$

Suppose the monthly risk-free interest rate $r_f = 0.003$. According to the mean-risk selection idea, we have the following model:

$$\begin{cases} \max E[\xi_1 x_1 + \xi_2 x_2 + \xi_3 x_3 + \xi_4 x_4 + \xi_5 x_5 + \xi_6 x_6] \\ \text{subject to:} \\ \qquad R(x_1, x_2, \cdots, x_6; r) \leq \alpha(r), \ \forall r \geq 0 \\ \qquad x_1 + x_2 + x_3 + x_4 + x_5 + x_6 = 1 \\ \qquad x_1, x_2, x_3, x_4, x_5, x_6 \geq 0 \end{cases} \tag{2.17}$$

where $\xi_1, \xi_2, \xi_3, \xi_4, \xi_5, \xi_6$ represent the random monthly returns of securities Hundsun (600570), Tianjin (600821), Wanwei (600063), Sany (600031), Baosteel (600019), and Tianchuang (600791), respectively, and

$$R(x_1, x_2, \cdots, x_6; r) = \Pr\{0.003 - (x_1\xi_1 + x_2\xi_2 + \cdots + x_6\xi_6) \geq r\}.$$

When solving Model (2.17), in order to check if the risk curve of a portfolio is below the investors' confidence curve, theoretically, we need to calculate the probability values of

$$\Pr\{0.003 - (x_1\xi_1 + x_2\xi_2 + \cdots + x_6\xi_6) \geq r\}$$

for any $r \geq 0$. However, since the risk curve is a continuous and decreasing function of the parameter r, in reality, we just need to calculate some probability values for finite r values in an certain interval by analyzing the confidence curve. In this example, the confidence curve is a horizontal line when $r \geq 0.3$. Since the risk curve is a decreasing function of r, risk curve will be below the confidence curve if $R(x_1, x_2, \cdots, x_6; r) \leq \alpha(r)$ holds for any

Table 2.6 Monthly Returns of Six Securities in Shanghai Stock Market

Month	600570	600821	600063	600031	600019	600791
01/2006	0.1137	0.0743	0.0650	-0.0288	-0.0080	0.0778
02/2006	-0.0616	0.0727	0.0727	-0.0608	0.0647	-0.2272
03/2006	-0.0413	0.1516	-0.0244	0.1910	-0.0329	0.0467
04/2006	0.0705	0.1625	-0.2000	0.1897	0.0026	-0.0105
05/2006	0.5082	0.2867	0.1875	0.2790	0.1279	0.1459
06/2006	-0.1830	-0.0674	0.0994	0.3116	-0.0185	0.1944
07/2006	-0.1098	0.2871	0.0745	-0.1950	-0.0708	-0.0252
08/2006	0.1833	-0.4041	-0.0074	0.0938	0.0228	0.0915
09/2006	-0.0042	-0.0445	0.0274	0.2865	0.0025	0.0528
10/2006	0.1188	-0.0740	-0.0777	0.1653	0.1757	-0.1176
11/2006	-0.0126	-0.0325	-0.0342	0.1186	0.4021	0.1059
12/2006	0.1882	0.0122	0.4986	0.5168	0.2658	0.0408
01/2007	0.4170	0.2931	0.6418	0.1179	0.1281	0.1363
02/2007	0.0715	0.3084	0.0598	-0.0421	-0.0305	0.1139
03/2007	0.3463	0.3429	0.1797	0.1789	0.0456	0.4603
04/2007	0.1065	0.3763	0.7050	-0.2355	0.1286	0.2774
05/2007	-0.0648	-0.0473	-0.0836	0.1474	0.1048	0.0339
06/2007	-0.2343	-0.3529	0.3543	0.2463	-0.0849	-0.3712
07/2007	0.2814	0.3292	-0.0808	0.0305	0.2264	0.4440
08/2007	0.7710	0.5283	-0.0255	0.1455	0.3706	0.3005
09/2007	0.0704	0.0231	0.2963	-0.0222	-0.0162	0.4072
10/2007	-0.1126	-0.2406	-0.2678	0.2517	0.0176	-0.2360
11/2007	-0.1728	-0.1847	-0.2329	-0.3124	-0.2156	-0.2694
12/2007	0.2515	0.3057	0.3248	0.2981	0.2011	0.1580

$r \in [0, 0.3]$. Since risk curve is a continuous function of r, it is enough for us to set $r = 0$, $r = 0.02$, $r = 0.04$, $r = 0.06, \cdots, r = 0.3$ and check if the points on the risk curve are all lower than the points on the confidence curve. That is, we just need to solve the following model

$$\begin{cases} \max E[\xi_1 x_1 + \xi_2 x_2 + \xi_3 x_3 + \xi_4 x_4 + \xi_5 x_5 + \xi_6 x_6] \\ \text{subject to:} \\ \quad \Pr\{0.003 - (x_1\xi_1 + x_2\xi_2 + \cdots + x_6\xi_6) \geq 0\} \leq 0.25 \\ \quad \Pr\{0.003 - (x_1\xi_1 + x_2\xi_2 + \cdots + x_6\xi_6) \geq 0.02\} \leq 0.225 \\ \quad \Pr\{0.003 - (x_1\xi_1 + x_2\xi_2 + \cdots + x_6\xi_6) \geq 0.04\} \leq 0.2 \\ \quad \cdots \\ \quad \Pr\{0.003 - (x_1\xi_1 + x_2\xi_2 + \cdots + x_6\xi_6) \geq 0.12\} \leq 0.10 \\ \quad \Pr\{0.003 - (x_1\xi_1 + x_2\xi_2 + \cdots + x_6\xi_6) \geq 0.14\} \leq 0.09 \\ \quad \cdots \\ \quad \Pr\{0.003 - (x_1\xi_1 + x_2\xi_2 + \cdots + x_6\xi_6) \geq 0.3\} \leq 0.01 \\ \quad x_1 + x_2 + x_3 + x_4 + x_5 + x_6 = 1 \\ \quad x_1, x_2, x_3, x_4, x_5, x_6 \geq 0. \end{cases} \tag{2.18}$$

Since the probabilistic portfolio theory, for the most part, assumes that security returns are normally distributed, we assume in the example that the portfolio returns are normal random variables. Thus, to solve Model (2.18), the expected value and variance value of the portfolio return, i.e., $E[\xi_1 x_1 + \xi_2 x_2 + \xi_3 x_3 + \xi_4 x_4 + \xi_5 x_5 + \xi_6 x_6]$ and $V[\xi_1 x_1 + \xi_2 x_2 + \xi_3 x_3 + \xi_4 x_4 + \xi_5 x_5 + \xi_6 x_6]$, are first calculated according to the data provided in Table 2.6 (the calculation of expected value and variance value will be illustrated in the later section 2.5.2). Then, click "Insert" in the menu of Microsoft Excel and choose the command "Function". From the command "Function", choose the function "NORMDIST". By using "NORMDIST", given a r value, the probability value $\Pr\{\xi_1 x_1 + \xi_2 x_2 + \cdots + \xi_6 x_6 \leq 0.003 - r\}$ can be obtained. A run of "Solver" in the menu "Tool" of Microsoft Excel shows that in order to obtain the maximum expected return from the safe portfolios whose risk curves are totally below the confidence curve, the investors should assign their money according to Table 2.7. The maximum expected return is 0.1099.

The confidence curve $\alpha(r)$ and the risk curve $R(r)$ of the selected portfolio are drawn in Fig. 2.16. With the risk curve, each likely loss and the

Table 2.7 Allocation of Money to Six Securities

600570	600821	600063	600031	600019	600791
5.47%	0.00 %	21.03%	73.5%	0.00%	0.00%

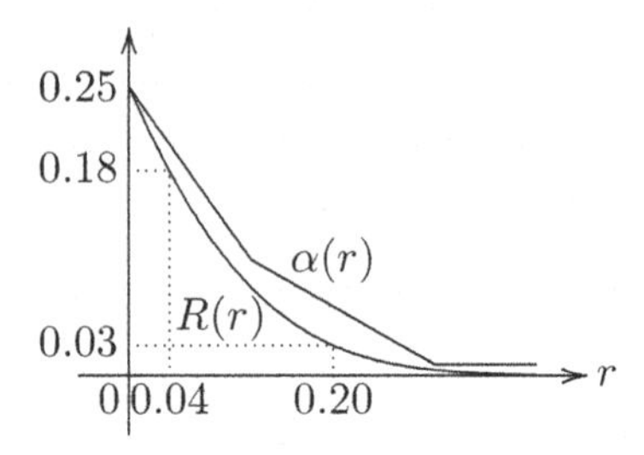

Fig. 2.16 Risk curve $R(r)$ and confidence curve $\alpha(r)$ of Model (2.17).

corresponding occurrence probability is observable. For example, at probability 18%, the maximal loss level is 0.04; and at probability 3%, the maximal loss level is 0.20. It can be seen that the risk curve of the selected portfolio is totally below the confidence curve. Given any loss value r, the loss occurrence probability is less than the investors' tolerable probability. Or given any occurrence probability $\alpha(r)$, the loss level is less than the investors' tolerance.

2.3 β-Return-Risk Model

2.3.1 β-Return-Risk Model

In the mean-risk model, expected return is used as the representative value of variable return. However, expected return provides only an average information rather than a specific return. Sometimes the investors like to pursue a specific target return instead of an average information. Since each investor knows that a portfolio return is variable, he/she must have some tolerance to inability to obtain the target return. However, at a preset probability level, the target return should be reached. To reflect this idea, we replace the expected return in the mean-risk model by the β-return and give the β-return-risk model as follows:

$$\begin{cases} \max \bar{f} \\ \text{subject to:} \\ \qquad \Pr\left\{\xi_1 x_1 + \xi_2 x_2 + \cdots + \xi_n x_n \geq \bar{f}\right\} \geq \beta \\ \qquad R(x_1, x_2, \cdots, x_n; r) \leq \alpha(r),\ \forall r \geq 0 \\ \qquad x_1 + x_2 + \cdots + x_n = 1 \\ \qquad x_i \geq 0, \quad i = 1, 2, \cdots, n \end{cases} \tag{2.19}$$

where β is the pre-set confidence level, and $\bar{f}$ the β-return defined as

$$\beta\text{-return} = \sup\{\bar{f} \mid \Pr\left\{\xi_1 x_1 + \xi_2 x_2 + \cdots + \xi_n x_n \geq \bar{f}\right\} \geq \beta\}$$

which means the maximal investment return the investors can obtain at confidence level β.

Since portfolio return is quite variable, investors may obtain a high return level as well as a low return level. In Model (2.19), investors in fact divide all the likely portfolio returns into two groups. One group includes all the likely "bad returns" which are lower than the risk-free interest rate r_f level. For returns in this group, the investors are cautious to each likely "bad return" and require that the occurrence probability of each "bad return" should be within the investors' tolerance level. Another group includes all the likely "good returns" which are expected to be higher than the risk-free interest rate r_f level. For returns in this group, the investors are sensitive to a specific target return which is expected to occur at a preset high confidence level. They want to pursue the maximum target return at the probability not less than the preset high confidence level.

Mathematically, the β-return-risk model is a maxmax model because it can also be expressed as follows:

$$\begin{cases} \max\limits_{x_1,x_2,\cdots,x_n} \quad \max\limits_{\bar{f}} \bar{f} \\ \text{subject to:} \\ \qquad \Pr\left\{\xi_1 x_1 + \xi_2 x_2 + \cdots + \xi_n x_n \geq \bar{f}\right\} \geq \beta \\ \qquad R(x_1, x_2, \cdots, x_n; r) \leq \alpha(r), \ \forall r \geq 0 \\ \qquad x_1 + x_2 + \cdots + x_n = 1 \\ \qquad x_i \geq 0, \quad i = 1, 2, \cdots, n, \end{cases} \tag{2.20}$$

in which $\max\limits_{\bar{f}} \bar{f}$ is the β-return.

2.3.2 Application Example

Again, we select an optimal portfolio from the six securities whose monthly returns are given in Table 2.6. The risk-free interest rate is still supposed to be $r_f = 0.003$, and the investors' confidence curve is still

$$\alpha(r) = \begin{cases} -1.25r + 0.25, & \text{when} \quad 0 \leq r \leq 0.12 \\ -0.5r + 0.16, & \text{when} \quad 0.12 \leq r \leq 0.3 \\ 0.01, & \text{when} \quad r \geq 0.3. \end{cases}$$

This time, the investors want to pursue the maximum specific return at probability not less than 70% from the safe portfolios. According to β-return-risk selection idea, we have the following model:

$$\begin{cases} \max \bar{f} \\ \text{subject to:} \\ \qquad \Pr\{\xi_1 x_1 + \xi_2 x_2 + \xi_3 x_3 + \xi_4 x_4 + \xi_5 x_5 + \xi_6 x_6 \geq \bar{f}\} \geq 0.7 \\ \qquad R(x_1, x_2, \cdots, x_6; r) \leq \alpha(r), \ \forall r \geq 0 \\ \qquad x_1 + x_2 + x_3 + x_4 + x_5 + x_6 = 1 \\ \qquad x_1, x_2, x_3, x_4, x_5, x_6 \geq 0 \end{cases} \tag{2.21}$$

where $\xi_1, \xi_2, \xi_3, \xi_4, \xi_5, \xi_6$ represent the random monthly returns of securities Hundsun(600570), Tianjin(600821), Wanwei(600019), Sany(600031), Baosteel(600063), and Tianchuang(600791), respectively, and $R(x_1, x_2, \cdots, x_6; r)$ the risk curve defined as

$$\Pr\{0.003 - (\xi_1 x_1 + \xi_2 x_2 + \xi_3 x_3 + \xi_4 x_4 + \xi_5 x_5 + \xi_6 x_6) \geq r\}.$$

By running "Solver" in "Excel", in order to obtain the maximum 0.7-return among the safe portfolios, the investors should allocate their money according to Table 2.8. The maximum return which can be obtained at probability 70% is 0.0341. The risk curve $R(r)$ and the confidence curve $\alpha(r)$ are drawn in Fig. 2.17

Table 2.8 Allocation of Money to Six Securities

600570	600821	600063	600031	600019	600791
5.58%	19.93 %	18.68%	53.33%	2.48%	0.00%

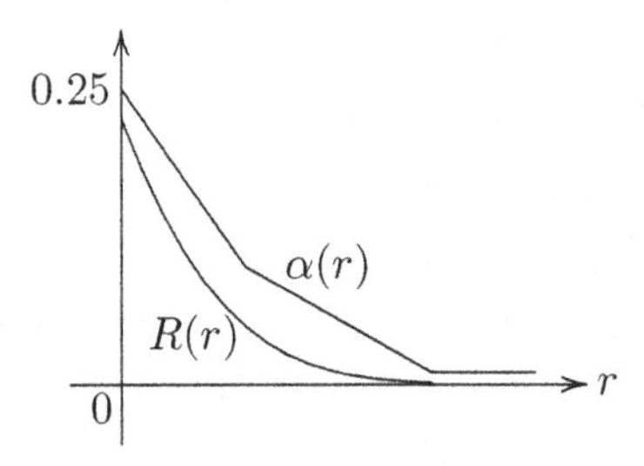

Fig. 2.17 Risk curve $R(r)$ and confidence curve $\alpha(r)$ of Model (2.21).

Remark 2.7. It is seen from application examples of mean-risk and β-return-risk models that even when risk-free interest rate, the alternative individual securities and the investors' confidence curve are same, adopting different selection criteria produces different results. The optimal portfolio and the corresponding expected return and the β-return values are shown in Table 2.9.

Table 2.9 Optimal Portfolios Produced by Different Selection Criteria

Optimal Portfolio	Mean-Risk Criterion	β-Return-Risk Criterion
$\xi_1(600570)$	5.47 %	5.58%
$\xi_2(600821)$	0.00 %	19.93%
$\xi_3(600063)$	21.03%	18.68%
$\xi_4(600031)$	73.50 %	53.33%
$\xi_5(600019)$	0.00 %	2.48%
$\xi_6(600791)$	0.00 %	0.00%
Expected Return	10.99%	10.44%
0.7-Return	2.68%	3.41%

2.4 Probability Minimization Model

2.4.1 Probability Minimization Model

In his safety first principle, Roy [83] proposed an alternative definition of risk to be the probability of the portfolio return below a sensitive disaster level. If the investors adopt this definition of risk, then the portfolio is optimal if the occurrence probability of portfolio return equal to or less than the disaster level is minimized. Let ξ_i be the i-th security returns and x_i the investment proportions, $i = 1, 2, \cdots, n$, respectively. The idea is expressed mathematically by probability minimization model as follows:

$$\left\{\begin{array}{l} \min \Pr\{\xi_1 x_1 + \xi_2 x_2 + \cdots + \xi_n x_n \leq d\} \\ \text{subject to:} \\ \qquad x_1 + x_2 + \cdots + x_n = 1 \\ \qquad x_i \geq 0, \quad i = 1, 2, \cdots, n \end{array}\right. \tag{2.22}$$

where d is the concerned disaster level.

Let us recall the definition of risk curve. The curve

$$R(x_1, x_2, \cdots, x_n; r) = \Pr\{r_f - (\xi_1 x_1 + \xi_2 x_2 + \cdots + \xi_n x_n) \geq r\}, \forall r \geq 0$$

is called the risk curve of the portfolio, where r_f is the risk-free interest rate. If r degenerates to one specific number r_0, then the risk curve becomes

$$\begin{aligned} R(x_1, x_2, \cdots, x_n; r_0) &= \Pr\{r_f - (\xi_1 x_1 + \xi_2 x_2 + \cdots + \xi_n x_n) \geq r_0\} \\ &= \Pr\{\xi_1 x_1 + \xi_2 x_2 + \cdots + \xi_n x_n \leq r_f - r_0\} \end{aligned}$$

which is just the one what Roy proposed. It is clear that $r_f - r_0 = d$.

If the investors pre-give a small probability value, then selection idea of Model (2.22) can also be expressed by the following model:

$$\begin{cases} \max \bar{d} \\ \text{subject to:} \\ \qquad \Pr\left\{\xi_1 x_1 + \xi_2 x_2 + \cdots + \xi_n x_n \le \bar{d}\right\} \le \alpha \\ \qquad x_1 + x_2 + \cdots + x_n = 1 \\ \qquad x_i \ge 0, \quad i = 1, 2, \cdots, n \end{cases} \tag{2.23}$$

where α is the pre-set small probability value and $\bar{d}$ the maximum low return at the preset small probability level defined as

$$\max\left\{\bar{d} \mid \Pr\left\{\xi_1 x_1 + \xi_2 x_2 + \cdots + \xi_n x_n \le \bar{d}\right\} \le \alpha\right\}.$$

Since probability measure is self-dual, the formula

$$\Pr\left\{\xi_1 x_1 + \xi_2 x_2 + \cdots + \xi_n x_n \le \bar{d}\right\} \le \alpha$$

is equivalent to

$$\Pr\left\{\xi_1 x_1 + \xi_2 x_2 + \cdots + \xi_n x_n \ge \bar{d}\right\} \ge 1 - \alpha,$$

which means that at the (high) probability value $1 - \alpha$, the portfolio return will be equal to or greater than this low return level $\bar{d}$. Or in other words, at the (high) probability value $1 - \alpha$, the portfolio loss will be equal to or less than $r_f - \bar{d}$ or $e - \bar{d}$ where e is the expected return of the portfolio. It is seen that $r_f - \bar{d}$ or $e - \bar{d}$ are different versions of Value-at-Risk (VaR).

For example, assume $r_f = 0.003$. If we set $\alpha = 5\%$ and get $\bar{d} = -0.03$, this means that at probability 95%, the portfolio return will be equal to or greater than -0.03 and the portfolio loss will be equal to or less than $0.003 - (-0.03) = 0.033$.

2.4.2 Application Example

Suppose the investors want to select the portfolio from the six securities whose monthly returns are given in Table 2.6. This time, the investors adopt the probability minimization criterion and set the disastrous return level at 0.022. Then the optimal portfolio is obtained via the following model:

$$\begin{cases} \min \Pr\left\{\xi_1 x_1 + \xi_2 x_2 + \xi_3 x_3 + \xi_4 x_4 + \xi_5 x_5 + \xi_6 x_6 \le 0.022\right\} \\ \text{subject to:} \\ \qquad x_1 + x_2 + x_3 + x_4 + x_5 + x_6 = 1 \\ \qquad x_i \ge 0, \quad i = 1, 2, \cdots, 6. \end{cases} \tag{2.24}$$

Based on the traditional assumption that the portfolio return is normally distributed, we use "NORMDIST" in "Function" in "Insert" in the menu of

Table 2.10 Allocation of Money to Six Securities

600570	600821	600063	600031	600019	600791
2.09%	20.21%	17.55%	49.34%	10.81%	0.00%

Microsoft Excel to calculate the objective function. Then, a run of the command "Solver" in "Tool" in Microsoft Excel shows that in order to minimize the probability of the portfolio return not greater than 0.022, the investors should allocate their money according to Table 2.10. The corresponding probability is 26.87%.

If the investors preset a probability 95%, and want to minimize the loss at this confidence, then according to the selection idea given in Model (2.23), we build the model as follows:

$$\left\{\begin{array}{l} \max \bar{d} \\ \text{subject to:} \\ \qquad \Pr\left\{\xi_1 x_1 + \xi_2 x_2 + \cdots + \xi_6 x_6 \leq \bar{d}\right\} \leq 5\% \\ \qquad x_1 + x_2 + \cdots + x_n = 1 \\ \qquad x_i \geq 0, \quad i = 1, 2, \cdots, 6. \end{array}\right. \tag{2.25}$$

A run of the "Solver" in "Excel" shows that the maximal $\bar{d} = -0.1043$, and the investors should allocate their money according to Table 2.11. In other words, at probability 95%, the maximal loss will be $0.003 - (-0.1043) = 0.1073$.

Table 2.11 Allocation of Money to Six Securities

600570	600821	600063	600031	600019	600791
0.00%	14.5%	13.73%	36.34%	29.72%	5.72%

2.5 Mean-Variance Model

The mean-variance model was first proposed by Markowitz [66]. The model opens the door for mathematical analysis on portfolios and serves as the foundation of modern finance theory. Till today, extensions and computation of the mean-variance model still remain a hot research topic.

2.5.1 Mean-Variance Model

According to Markowitz [66, 67], the return of a portfolio was represented by the mean and the risk was quantified by the variance. Then the investors should strike a balance between maximizing the return and minimizing the risk. In the case of maximizing the return at a given specific level of risk, the standard formulation of Markowitz model is as follows:

$$\begin{cases} \max E[x_1\xi_1 + x_2\xi_2 + \cdots + x_n\xi_n] \\ \text{subject to:} \\ \qquad V[x_1\xi_1 + x_2\xi_2 + \cdots + x_n\xi_n] \le \gamma \\ \qquad x_1 + x_2 + \cdots + x_n = 1 \\ \qquad x_i \ge 0, \quad i = 1, 2, \cdots, n \end{cases} \tag{2.26}$$

where E denotes the expected value operator, V the variance operator, x_i the investment proportions in securities i, ξ_i the random returns of the i-th securities, $i = 1, 2, \cdots, n$, respectively, and γ is the maximum variance level the investors can tolerate.

In the case of minimizing the risk for a given level of return, the formulation is as follows:

$$\begin{cases} \min V[x_1\xi_1 + x_2\xi_2 + \cdots + x_n\xi_n] \\ \text{subject to:} \\ \qquad E[x_1\xi_1 + x_2\xi_2 + \cdots + x_n\xi_n] \ge \alpha \\ \qquad x_1 + x_2 + \cdots + x_n = 1 \\ \qquad x_i \ge 0, \quad i = 1, 2, \cdots, n \end{cases} \tag{2.27}$$

where α represents the minimum expected return the investors can accept.

According to Markowitz, a portfolio is efficient if it is impossible to obtain higher expected return with no greater variance value, or it is impossible to obtain less variance value with no less expected return. All efficient portfolios make up the efficient frontier. An efficient portfolio is in fact a solution of the following optimization model with two objectives:

$$\begin{cases} \max E[x_1\xi_1 + x_2\xi_2 + \cdots + x_n\xi_n] \\ \min V[x_1\xi_1 + x_2\xi_2 + \cdots + x_n\xi_n] \\ \text{subject to:} \\ \qquad x_1 + x_2 + \cdots + x_n = 1 \\ \qquad x_i \ge 0, \quad i = 1, 2, \cdots, n. \end{cases} \tag{2.28}$$

Different investors will find different optimal portfolios from the efficient frontier according to their preferences to risk aversion, i.e., tradeoff of variance and expected return. The investors can pick their own portfolio from the efficient frontier according to their own utility function which evaluates the risk-return-tradeoff.

2.5.2 Application Example

Before illustrating how to use the mean-variance model, let us first show how to calculate the expected value and variance value of the portfolio return.

Expected Return of a Portfolio

According to the law of large number, we can use average value of N samples of portfolio returns to approximate the expected return of the portfolio. Let ξ_i represent the random returns of the i-th securities, r_{ij} the j-th sample returns of the i-th securities, and x_i the investment proportions in the i-th securities, $i = 1, 2, \cdots, n, j = 1, 2, \cdots, N$, respectively. Then we have

$$E[x_1\xi_1 + x_2\xi_2 + \cdots + x_n\xi_n] = \sum_{j=1}^{N}(x_1r_{1j} + x_2r_{2j} + \cdots + x_nr_{nj})/N.$$

For example, consider a case in which we have a portfolio BSW containing 20% Baosteel (600019), 20% Sany Heavy Industry (600031), and the rest 60% of Wanwei Updated High-Tech (600063). For convenience, we just use one year's data to illustrate the calculation of the monthly expected return of the portfolio. We denote the three security returns by ξ_1, ξ_2 and ξ_3, respectively, and use $r_{1j}, r_{2j}, r_{3j}, j = 1, 2, \cdots, 12$, to represent the j-th monthly returns of ξ_1, ξ_2 and ξ_3, respectively. The expected return of the portfolio BSW is calculated as follows (also see Table 2.12):

$$\begin{aligned}
&E[0.20\xi_1 + 0.20\xi_2 + 0.60\xi_3] \\
&= \sum_{j=1}^{12}(0.2r_{1j} + 0.2r_{2j} + 0.6r_{3j})/12 \\
&= 0.1216.
\end{aligned}$$

If we use "Excel", we can use the function "SUMPRODUCT" to easily get the 12 numbers of sample portfolio returns, i.e.,

$$R_j = 0.2r_{1j} + 0.2r_{2j} + 0.6r_{3j}, \quad j = 1, 2, \cdots, 12.$$

Then use the function "AVERAGE" to calculate the mean of the 12 numbers of sample portfolio returns R_j. We illustrate it in Table 2.12.

Variance Value of a Portfolio Return

Variance value can be obtained according to the definition of it, i.e.,

$$\begin{aligned}
&V[x_1\xi_1 + x_2\xi_2 + \cdots + x_n\xi_n] \\
&= E[(x_1\xi_1 + x_2\xi_2 + \cdots + x_n\xi_n - E[x_1\xi_1 + x_2\xi_2 + \cdots + x_n\xi_n])^2].
\end{aligned}$$

Table 2.12 Computation of Expected Return and Variance of Portfolio BSW

Month j	r_{1j}	r_{2j}	r_{3j}	$0.2r_{1j}+0.2r_{2j}+0.6r_{3j}$ (use SUMPRODUCT)
01/2007	0.1281	0.1179	0.6418	0.4343
02/2007	-0.0305	-0.0421	0.0598	0.0214
03/2007	0.0456	0.1789	0.1797	0.1527
04/2007	0.1286	-0.2355	0.7050	0.4016
05/2007	0.1048	0.1474	-0.0836	0.0003
06/2007	-0.0849	0.2463	0.3543	0.2449
07/2007	0.2264	0.0305	-0.0808	0.0029
08/2007	0.3706	0.1455	-0.0255	0.0880
09/2007	-0.0162	-0.0222	0.2963	0.1701
10/2007	0.0176	0.2517	-0.2678	-0.1068
11/2007	-0.2156	-0.3124	-0.2329	-0.2453
12/2007	0.2011	0.2981	0.3248	0.2947
Mean	(use AVERAGE)			0.1216
Variance	(use VAR)			0.0410

Thus the variance value of portfolio BSW is obtained as follows (also see Table 2.12):

$$\begin{aligned}
&V[x_1\xi_1+x_2\xi_2+x_3\xi_3]\\
&=E[(0.2\xi_1+0.2\xi_2+0.6\xi_3-E[0.2\xi_1+0.2\xi_2+0.6\xi_3])^2]\\
&=\sum_{j=1}^{12}\Big(0.2r_{1j}+0.2r_{2j}+0.6r_{3j}-0.1215\Big)^2\\
&=0.0376.
\end{aligned}$$

In Excel, we can also simply use the function "VAR" to get the variance value of twelve numbers of the sample portfolio returns $R_j=0.2r_{1j}+0.2r_{2j}+0.6r_{3j}, j=1,2,\cdots,12$. The calculation of variance is shown in Table 2.12. We notice that the variance value in Table 2.12 is somewhat different from the variance value obtained via the definition of variance. This is because the sample number is small in our case.

In some literature, variance value of a portfolio return is usually calculated via variance values of individual securities and the covariance values between two different individual security returns. That is, in portfolio BSW case, variance value is obtained via the formula

$$\begin{aligned}
&V[x_1\xi_1+x_2\xi_2+x_3\xi_3]\\
&=x_1^2V[\xi_1]+x_2^2V[\xi_2]+x_3^2V[\xi_3]\\
&\quad+2x_1x_2Cov(\xi_1,\xi_2)+2x_1x_3Cov(\xi_1,\xi_3)+2x_2x_3Cov(\xi_2,\xi_3)
\end{aligned}$$

where Cov is the covariance between two different random variables. We can see that using definition of variance to calculate the variance value is a much easier way, and can avoid much unnecessary computation work.

Examples of Mean-Variance Model

Suppose the investors want to select the portfolio from the securities Hundsun Electronics (600570), Tianjin Quanye (600821), Wanwei Updated High-Tech (600063), Sany Heavy Industry (600031), Baosteel (600019), and Tianchuang Property (600791). The monthly returns of the six securities are given in Table 2.6. Assume that the investors adopt the mean-variance selection principle and want to pursue the maximum expected return with the maximum tolerable variance at 0.015. Then the mean-variance selection model can be built as follows:

$$\begin{cases} \max E[x_1\xi_1 + x_2\xi_2 + x_3\xi_3 + x_4\xi_4 + x_5\xi_5 + x_6\xi_6] \\ \text{subject to:} \\ \qquad V[x_1\xi_1 + x_2\xi_2 + x_3\xi_3 + x_4\xi_4 + x_5\xi_5 + x_6\xi_6] \le 0.015 \\ \qquad x_1 + x_2 + x_3 + x_4 + x_5 + x_6 = 1 \\ \qquad x_1, x_2, x_3, x_4, x_5, x_6 \ge 0 \end{cases} \tag{2.29}$$

where $\xi_1, \xi_2, \xi_3, \xi_4, \xi_5, \xi_6$ represent the random returns of securities Hundsun (600570), Tianjin (600821), Wanwei (600063), Sany (600031), Baosteel (600019) and Tianchuang (600791), respectively, and E and V denote the expected value operator and variance operator, respectively.

We use the command "Solver" in menu "Tool" in Microsoft Excel to solve Model (2.29). When calculating the expected value and variance value of the portfolio, we use the method introduced above. The difference is that in the above introduction, the investment proportion in three securities are determinant, while when computing the expected value and variance value of the portfolio in Model (2.29), the decision variables $x_1, x_2, \cdots, x_6$ replace the determinant proportions. A run of the command "Solver" in "Tool" in Microsoft Excel shows that in order to obtain the maximum expected return with variance not greater than 0.015, the investors should assign their money according to Table 2.13. The maximum expected value is 0.097.

Table 2.13 Allocation of Money to Six Securities

600570	600821	600063	600031	600019	600791
0.00%	16.47%	14.87%	40.35%	24.22%	4.09%

If the investors want to minimize the variance value with the investment return not less than 0.10, then the mean-variance model is built as follows:

$$\begin{cases} \min V[x_1\xi_1 + x_2\xi_2 + x_3\xi_3 + x_4\xi_4 + x_5\xi_5 + x_6\xi_6] \\ \text{subject to:} \\ \qquad E[x_1\xi_1 + x_2\xi_2 + x_3\xi_3 + x_4\xi_4 + x_5\xi_5 + x_6\xi_6] \geq 0.10 \\ \qquad x_1 + x_2 + x_3 + x_4 + x_5 + x_6 = 1 \\ \qquad x_1, x_2, x_3, x_4, x_5, x_6 \geq 0. \end{cases} \tag{2.30}$$

We again use the command "Solver" in menu "Tool" in Microsoft Excel to solve Model (2.30). A run of the command "Solver" shows that in order to minimize the variance value with the expected return not less than 0.10, the investors should assign their money according to Table 2.14. The minimum variance value is 0.016.

Table 2.14 Allocation of Money to Six Securities

600570	600821	600063	600031	600019	600791
0.00%	19.82%	16.80%	47.15%	14.88%	1.35%

By changing the preset variance value in Model (2.29) or changing the preset expected return in Model (2.30), the efficient frontier of the alternative portfolios is obtained and drawn in Fig. 2.18.

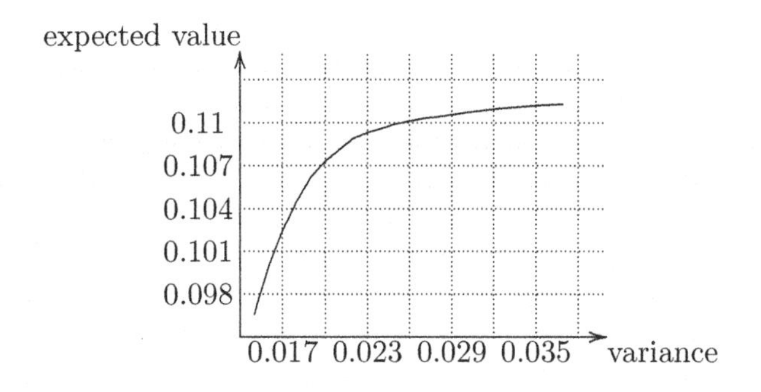

Fig. 2.18 Efficient frontier for alternative portfolios in the mean-variance application example.

2.5.3 Mean-Semivariance Model

When probability distributions of security returns are asymmetric, variance becomes a deficient measure of investment risk because the selected portfolio based on variance may have a potential danger to sacrifice too much expected return in eliminating both low and high return extremes. For example, suppose we have a security whose expected return is high but whose variance of return is also very high because the security has very great positive deviations from the expected return. In Model (2.26) in page 40, since the constraint is that the variance value of the portfolio return must not be higher than the preset level, then it is very likely that this security with high expected return and high positive deviations will be deleted, yet this security is what we like. In Model (2.27) in page 40, since the objective is that the variance of the portfolio return should be minimized, then it is very likely that this security with high variance will be deleted though positive deviation and high expected return are what we welcome. To overcome the limitation of the mean-variance models, semivariance [67] was proposed to replace variance as the measure of risk. Semivariance separates undesirable downside fluctuations of security returns from the desirable upside fluctuations and only pays attention to returns falling below the expected return. Therefore, it matches investors' notion about risk and gains popularity among investors. Then , in case when the investors can give a tolerable level of risk, they should maximize the expected return at the given level of semivariance. Thus, we have the mean-semivariance model as follows:

$$\begin{cases} \max E[x_1\xi_1 + x_2\xi_2 + \cdots + x_n\xi_n] \\ \text{subject to:} \\ \qquad SV[x_1\xi_1 + x_2\xi_2 + \cdots + x_n\xi_n] \le \gamma \\ \qquad x_1 + x_2 + \cdots + x_n = 1 \\ \qquad x_i \ge 0, \quad i = 1, 2, \cdots, n \end{cases} \tag{2.31}$$

where γ denotes the maximum semivariance level the investors can tolerate, E the expected value operator, and SV the semivariance of the random variables.

When the investors preset an expected return level that they feel satisfactory and want to minimize the risk at this given level of expected return, the optimization model becomes:

$$\begin{cases} \min SV[x_1\xi_1 + x_2\xi_2 + \cdots + x_n\xi_n] \\ \text{subject to:} \\ \qquad E[x_1\xi_1 + x_2\xi_2 + \cdots + x_n\xi_n] \ge \alpha \\ \qquad x_1 + x_2 + \cdots + x_n = 1 \\ \qquad x_i \ge 0, \quad i = 1, 2, \cdots, n \end{cases} \tag{2.32}$$

where α denotes the minimum expected investment return that the investors can accept.

Similarly, as in mean-variance model, a portfolio is efficient if it is impossible to obtain higher expected return with no greater semivariance value, or it is impossible to obtain less semivariance value with no less expected return. The efficient portfolio is in fact the solution of the following optimization model with two objectives:

$$\begin{cases} \max E[x_1\xi_1 + x_2\xi_2 + \cdots + x_n\xi_n] \\ \min SV[x_1\xi_1 + x_2\xi_2 + \cdots + x_n\xi_n] \\ \text{subject to:} \\ \qquad x_1 + x_2 + \cdots + x_n = 1 \\ \qquad x_i \geq 0, \quad i = 1, 2, \cdots, n. \end{cases} \tag{2.33}$$

Many scholars have developed computation algorithms for the mean-semivariance model, for example, the numerical algorithm by Mao [65], a viable computational scheme for generating mean-semivariance efficient portfolios by Hogan and Warren [23], a semi-linear model algorithm to approximate the mean-semivariance model by Ang [2], a simple approximation method for semivariance by Choobineh and Branting [9], and the critical line algorithm to compute mean-semivariance efficient sets by Markowitz [68].

2.6 Hybrid Intelligent Algorithm

Sometimes it is difficult to use traditional methods to solve the portfolio selection problems. For example, in the application example of mean-risk model in Section 2.2.4, if some security returns are normally distributed and the other security returns are lognormally distributed, then it is hard to use Excel to solve the problem. Therefore, we introduce a hybrid intelligent algorithm as the general solution algorithm for probabilistic portfolio selection problems. Generally speaking, we employ stochastic simulation to calculate the objective and constraint values first. Then we embed the simulation results into the genetic algorithm and use the genetic algorithm to find the optimal solution. In order to employ stochastic simulation, we need to know how to generate random numbers.

2.6.1 Random Number Generation

Random number generation is the first step in stochastic simulation. The methods of random number generation have already been discussed and documented in many books such as in Fishman [19], Law and Kelton [48], Bratley et al. [6], and so on. Since we mainly use random variables with uniform distribution, normal distribution, and lognormal distribution, respectively, we summarize here the computer methods for generating random numbers from the three types of distributions. We omit the explanation why the methods can produce the random numbers from the required distributions because it is out of our focus and is of independent interest. For detailed expositions on random number generation, the interested readers can refer to the books mentioned above.

Uniform Distribution
Recall that a random variable ξ, denoted by $\mathcal{U}(a,b)$, has a uniform distribution if its probability density function is defined by

$$f(t) = \begin{cases} \dfrac{1}{b-a}, & \text{if } a \le t \le b \\ 0, & \text{otherwise.} \end{cases}$$

Generation of uniformly distributed random numbers is the basis of generating other types of random numbers. It is made by a deterministic sequence called *pseudorandom numbers* on a digital computer. The deterministic sequence is regarded to be random numbers because the numbers generated by the sequence are stochastically independent and uniformly distributed. The C library for any types of computers has provided the subfunction of generating pseudorandom numbers. This subfunction, which is defined as

```
♯ include ⟨ stdlib.h ⟩
int rand(void),
```

produces a peseudorandom integer between 0 and RAND_ MAX, where RAND_ MAX is defined as $2^{15}-1$ in stdlib.h. Thus a uniformly distributed random number in the interval $[a,b]$ can be obtained by the following steps:

Step 1. $u = \text{rand}(\,)$.
Step 2. $u \leftarrow u/\text{RAND\_MAX}$.
Step 3. Return $a + u(b-a)$.

Normal Distribution
Recall that a random variable ξ, denoted by $\mathcal{N}(\mu,\sigma^2)$, is normally distributed if it has the following probability density distribution

$$f(t) = \frac{1}{\sigma\sqrt{2\pi}} \exp\left[-\frac{(t-\mu)^2}{2\sigma^2}\right], \quad \sigma > 0, \quad t \in \Re,$$

where μ and σ^2 are the expected value and variance of the variable, respectively. It can be produced via the following steps:

Step 1. Generate μ_1 and μ_2 from $\mathcal{U}(0,1)$.
Step 2. $y = [-2\ln(\mu_1)]^{\frac{1}{2}} \sin(2\pi\mu_2)$.
Step 3. Return $\mu + \sigma y$.

Lognormal Distribution
Recall that if t is from $\mathcal{N}(\mu,\sigma^2)$, then the variable $y = \exp[t]$ is lognormally distributed with the probability density distribution

$$f(y) = \frac{1}{y\sigma\sqrt{2\pi}} \exp\left[-\frac{(\ln y - \mu)^2}{2\sigma^2}\right], \quad \sigma > 0, \quad y > 0.$$

It is denoted by $\ln \mathcal{N}(\mu, \sigma^2)$, and can be produced via the following steps:

Step 1. Generate t from $\mathcal{N}(\mu, \sigma^2)$.
Step 2. Return $\exp[t]$.

2.6.2 Stochastic Simulations

Since the variance value and the semivariance value are a kind of expected value, essentially, we need to calculate the values of three types of uncertain functions in probabilistic portfolio selection models. For expression convenience, let investment proportions $\boldsymbol{x} = (x_1, x_2, \cdots, x_n)$ and random security returns $\boldsymbol{\xi} = (\xi_1, \xi_2, \cdots, \xi_n)$. The three types of uncertain functions are as follows:

$$\text{Expected Value:} \quad E[\boldsymbol{x\xi}],$$

$$\text{Probability Value:} \quad \Pr\{\boldsymbol{x\xi} \le r\},$$

$$\beta\text{-Return:} \quad \max\{\bar{f} \mid \Pr\{\boldsymbol{x\xi} \ge \bar{f}\} \ge \beta\}.$$

Simulation for Expected Value $E[\boldsymbol{x\xi}]$

In order to compute *Expected Value*, we N times generate the random vector $\boldsymbol{a}_j = (a_{j1}, a_{j2}, \cdots, a_{jn})$ from the corresponding probability distributions of random variables $\xi_1, \xi_2, \cdots, \xi_n$, respectively, where j represents the number of generation times of the random vector. It follows from the strong law of large numbers that

$$\frac{\sum_{j=1}^{N} \boldsymbol{x a}_j}{N} \longrightarrow E[\boldsymbol{x\xi}], \quad \text{almost surely} \tag{2.34}$$

as $N \to \infty$. Therefore, the expected value $E[\boldsymbol{x\xi}]$ can be calculated approximately by $\frac{1}{N}\sum_{j=1}^{N} \boldsymbol{x}_j \boldsymbol{a}_j$ as long as the number N is big enough. Therefore, the simulation procedures are designed as follows:

Step 1. Set $e = 0$.
Step 2. Generate the random vector $\boldsymbol{a} = (a_1, a_2, \cdots, a_n)$ from the corresponding probability distributions of random variables $\xi_1, \xi_2, \cdots, \xi_n$, respectively.
Step 3. $e \leftarrow e + \boldsymbol{xa}$.

Step 4. Repeat the second and third steps N times where N is a sufficiently large integer number.
Step 5. $E[\boldsymbol{x\xi}] = e/N$.

Simulation for Probability value $\Pr\{\boldsymbol{x\xi} \le r\}$:

To compute the probability value $\Pr\{\boldsymbol{x\xi} \le r\}$, we generate the random vector $\boldsymbol{a}_j = (a_{j1}, a_{j2}, \cdots, a_{jn})$ from the corresponding probability distributions of random variables $\xi_1, \xi_2, \cdots, \xi_n$, respectively, where $j = 1, 2, 3, \cdots$, represent the generation times of the random vector. Equivalently, this means we generate ω_j from Ω according to the corresponding probability measure Pr and produce $\boldsymbol{\xi}(\omega_j)$ for $j = 1, 2, 3, \cdots$, times. Let N' represent the number of occasions that the product value of $\boldsymbol{x}$ and $\boldsymbol{\xi}(\omega_j)$ not greater than the value r, i.e., $\boldsymbol{x\xi}(\omega_j) \le r$, happens. Let us define

$$f(\boldsymbol{x}, \boldsymbol{\xi}(\omega_j)) = \begin{cases} 1, & \text{if } \boldsymbol{x\xi}(\omega_j) \le r, \\ 0, & \text{otherwise.} \end{cases}$$

Then $E[f(\boldsymbol{x}, \boldsymbol{\xi}(\omega_j))] = \Pr\{\boldsymbol{x\xi}(\omega_j) \le r\}$ for all j's, and for $j = 1, 2, \cdots, N$, we have $N' = \sum_{j=1}^{N} f(\boldsymbol{x}, \boldsymbol{\xi}(\omega_j))$. According to the strong law of large numbers, we have

$$\frac{N'}{N} = \frac{\sum_{j=1}^{N} f(\boldsymbol{x}, \boldsymbol{\xi}(\omega_j))}{N} \longrightarrow E[f(\boldsymbol{x}, \boldsymbol{\xi})] = \Pr\{\boldsymbol{x\xi} \le r\}, \quad \text{almost surely}$$

as $N \to \infty$. Therefore, the probability value $\Pr\{\boldsymbol{x\xi} \le r\}$ can be calculated approximately by N'/N as long as the number N is big enough. The simulation procedures are designed as follows:

Step 1. Set $N' = 0$.
Step 2. Generate the random vector $\boldsymbol{a} = (a_1, a_2, \cdots, a_n)$ from the corresponding probability distributions of random variables $\xi_1, \xi_2, \cdots, \xi_n$, respectively.
Step 3. $N' \leftarrow N' + 1$ if $\boldsymbol{xa} \le r$.
Step 4. Repeat the second and third steps N times where N is a sufficiently large integer number.
Step 5. $\Pr\{\boldsymbol{x\xi} \le r\} = N'/N$.

Simulation for β-Return value $\max\{\bar{f} \mid \Pr\{\boldsymbol{x\xi} \ge \bar{f}\} \ge \beta\}$:

When $\boldsymbol{\xi}$ is a continuous random vector, the β-return, i.e., the maximum $\bar{f}$, is achieved at the equality case

$$\Pr\{\boldsymbol{x\xi} \ge \bar{f}\} = \beta.$$

In order to compute the β-Return value, we generate the random vector $\boldsymbol{a}_j = (a_{j1}, a_{j2}, \cdots, a_{jn})$ from the corresponding probability distributions of random variables $\xi_1, \xi_2, \cdots, \xi_n$, respectively, where $j = 1, 2, 3, \cdots$, represent the generation times of the random vector. Equivalently, this means we generate ω_j from Ω according to the corresponding probability measure Pr and produce $\boldsymbol{\xi}(\omega_j)$ for $j = 1, 2, 3, \cdots$, times. Let us define

$$f(\boldsymbol{x}, \boldsymbol{\xi}(\omega_j)) = \begin{cases} 1, & \text{if } \boldsymbol{x}\boldsymbol{\xi}(\omega_j) \geq \overline{f} \\ 0, & \text{otherwise.} \end{cases}$$

Then we have $E[f(\boldsymbol{x}, \boldsymbol{\xi}(\omega_j))] = \beta$ for all j's. For $j = 1, 2, \cdots, N$, according to the strong law of large numbers, we have

$$\frac{\sum\limits_{j=1}^{N} f(\boldsymbol{x}, \boldsymbol{\xi}(\omega_j))}{N} \longrightarrow E[f(\boldsymbol{x}, \boldsymbol{\xi})] = \beta, \quad \text{almost surely}$$

as $N \to \infty$. Since the sum $\sum_{j=1}^{N} f(\boldsymbol{x}, \boldsymbol{\xi}(\omega_j))$ is just the number of $\boldsymbol{\xi}(\omega_j)$ satisfying $\boldsymbol{x}\boldsymbol{\xi}(\omega_j)) \geq \overline{f}$ for $j = 1, 2, \cdots, N$, then $\overline{f}$ is just the N'th largest element in the sequence $\{\boldsymbol{x}\boldsymbol{\xi}(\omega_1), \boldsymbol{x}\boldsymbol{\xi}(\omega_2), \cdots, \boldsymbol{x}\boldsymbol{\xi}(\omega_N)\}$, where N' is the integer part of βN. Therefore, the simulation procedures are designed as follows:

Step 1. Generate N times the random vector $\boldsymbol{a}_j = (a_{j1}, a_{j2}, \cdots, a_{jn}), j = 1, 2, \cdots, N$, from the corresponding probability distributions of random variables $\xi_1, \xi_2, \cdots, \xi_n$, respectively, where N is a sufficiently big integer number.
Step 2. Set $f_j = \boldsymbol{x}\boldsymbol{a}_j$ for $j = 1, 2, \cdots, N$.
Step 3. Set $N^{'}$ as the integer part of βN.
Step 4. Return the $N^{'}$-th largest element in $\{f_1, f_2, \cdots, f_N\}$ as the approximation of the β-return value.

Example 2.16. Suppose we have a portfolio A which is composed half of the normally distributed security return $\xi_1 \sim \mathcal{N}(0.1, 0.04)$ and half of the uniformly distributed security return $\xi_2 \sim \mathcal{U}(-0.5, 1.2)$. A run of the simulation with 8000 cycles shows that the probability of the portfolio return not greater than 0.03 is 0.2736, i.e., $\Pr\{0.5\xi_1 + 0.5\xi_2 \leq 0.03\} = 0.2736$. The simulation procedures are as follows:

Step 1. Set $N^{'} = 0$.
Step 2. Generate random numbers a from the normal distribution $\mathcal{N}(0.1, 0.04)$ and b from the uniform distribution $\mathcal{U}(-0.5, 1.2)$.
Step 3. $N^{'} \leftarrow N^{'} + 1$ if $0.5a + 0.5b <= 0.03$.
Step 4. Repeat the second and third steps 8000 times.
Step 5. $\Pr\{0.5\xi_1 + 0.5\xi_2 <= 0.03\} = N'/8000$.

Example 2.17. Suppose we have another portfolio B which is composed of 40% of security 1 and 60% of security 2. The return of security 1 is regarded to be a normal random variable $\xi_1 \sim \mathcal{N}(0.1, 0.01)$, and the return of security 2 is believed to be a lognormal random variable $\xi_2 \sim \ln\mathcal{N}(0.15, 0.02) - 1$. A run of the simulation with 8000 cycles shows that the probability of the portfolio return not greater than 0.05 is 0.1976, i.e., $\Pr\{0.4\xi_1 + 0.6\xi_2 \leq 0.05\} = 0.1976$. The simulation procedures are as follows:

Step 1. Set $N^{'} = 0$.
Step 2. Generate random numbers a from the normal distribution $\mathcal{N}(0.1, 0.01)$ and b from the lognormal distribution $\ln\mathcal{N}(0.15, 0.02)$. Then let $b \leftarrow b - 1$.
Step 3. $N^{'} \leftarrow N^{'} + 1$ if $0.4a + 0.6b <= 0.05$.
Step 4. Repeat the second and third steps 8000 times.
Step 5. $\Pr\{0.4\xi_1 + 0.6\xi_2 <= 0.05\} = N'/8000$.

Example 2.18. For the above mentioned portfolio B, if we set the confidence level $\beta = 0.8$, a run of the simulation with 6000 cycles shows that the 0.8-return $\bar{f} = 0.0497$. The simulation procedures are as follows:

Step 1. Generate random numbers a from the normal distribution $\mathcal{N}(0.1, 0.01)$ and b from the lognormal distribution $\ln\mathcal{N}(0.15, 0.02)$. Then let $b \leftarrow b - 1$.
Step 2. Set $f_j = 0.4a + 0.6b$ for $j = 1, 2, \cdots, 6000$.
Step 3. Set $N^{'} = 0.8 \times 6000 = 4800$.
Step 4. Return the 4800-th largest element in $\{f_1, f_2, \cdots, f_{6000}\}$ as the approximation of the 0.8-return value.

Example 2.19. To calculate the variance value of the return of the above mentioned portfolio B, we need to know the expected return of the portfolio first. According to the result of Example 2.7 in page 21, the expected return of security 2 is $\exp(0.15{+}0.02/2){-}1 = \exp(0.16){-}1 = 0.1735$. From the linearity property of expected value (see Theorem 2.3 in page 15), the expected return of the portfolio is $0.4 \times 0.1 + 0.6 \times 0.1735 = 0.1441$. A run of the simulation with 8000 cycles shows that

$$V[0.4\xi_1 + 0.6\xi_2] = 0.0115.$$

The simulation procedures for calculating variance value are as follows:

Step 1. Set $e = 0$.
Step 2. Generate random numbers a from the normal distribution $\mathcal{N}(0.1, 0.01)$ and b from the lognormal distribution $\ln\mathcal{N}(0.15, 0.01)$. Then let $b \leftarrow b - 1$.
Step 3. $e \leftarrow e + (0.4a + 0.6b - 0.1441)^2$.
Step 4. Repeat the second and third steps 8000 times.
Step 5. $V\left[0.4\xi_1 + 0.6\xi_2\right] = e/8000$.

Example 2.20. For the above mentioned portfolio B, a run of the simulation with 8000 cycles shows that

$$SV[0.4\xi_1 + 0.6\xi_2] = 0.0054.$$

The simulation procedures for calculating the semivariance value are as follows:

Step 1. Set $e = 0$.
Step 2. Generate random numbers a from the normal distribution $\mathcal{N}(0.1, 0.01)$ and b from the lognormal distribution $\ln\mathcal{N}(0.15, 0.02)$. Then let $b \leftarrow b - 1$.
Step 3. Let $e \leftarrow e + (0.4a + 0.6b - 0.1441)^2$ if $0.4a + 0.6b - 0.1441 <= 0$.
Step 4. Repeat the second and third steps 8000 times.
Step 5. $SV\,[0.4\xi_1 + 0.6\xi_2] = e/8000$.

2.6.3 Genetic Algorithm

Genetic algorithms (GAs) were first proposed by Holland [24]. They are a stochastic search method that finds the optimal solution based on the natural mechanism of "survival of the fittest". GAs have successfully solved many complex industrial optimization problems that are difficult to solve by traditional methods since its introduction. By group searching and group exchanging, GAs are quite robust and can avoid getting stuck at a local optimal solution. In addition, GAs do not require the specific mathematical analysis of optimization problems, which makes GAs an easy-to-use method for the users who are not necessarily good at mathematics.

In GAs, there is an important term *chromosome* which is a genetic representation of a solution to the problem. The chromosome is not necessarily the solution itself. It can be a coding of a solution but must be able to be decoded to the solution. The predetermined integer numbers of the chromosomes form a *population*. This predetermined integer number is called *population size* or *pop_size* for brief. GAs have five basic components in general [71]: first, the chromosomes; second, the way to create an initial population of chromosomes; third, an evaluation function rating chromosomes in terms of fitness; fourth, crossover and mutation operators that alter the genetic composition of chromosomes during reproduction; fifth, values of parameters of GAs.

GAs begin with a *pop_size* number of randomly generated feasible chromosomes, the generation process of which is called *initialization*. Then the fitness of each chromosome is evaluated by the *evaluation function*. Based on the fitness of each chromosome, a new population will be formed by a *selection process* using a mechanism which is fitness proportional. That is, in the *selection process* the likelihood of the old chromosome being selected to enter into the new population is proportional to its fitness. Next, the new population of chromosomes undergoes *crossover* and *mutation* operations to produce the *offspring*. Crossover creates new chromosomes by combining the parts from two chromosomes, and mutation creates new chromosomes by making changes in a single chromosome. After crossover and mutation, the population enters a new *generation*, and the new rounds of selection, crossover and mutation will continue until the algorithm converges to the best chromosome or a given number of cycles is met. We take the best chromosome and decode it into the solution which is regarded to be the optimal solution of the optimization problem. Let $B(i)$ represent a population of chromosomes for generation i, and $C(i)$ the offspring produced in the generation i. The general procedures of the GAs are as follows:

General Procedures: Genetic Algorithms
begin
  $i \leftarrow 0$;
  initialize $B(i)$;
  evaluate $B(i)$;
  while *(termination condition not satisfied)* **do**
  begin
    crossover and mutate $B(i)$ *to produce* $C(i)$;
    evaluate $C(i)$;
    select $B(i+1)$ *from* $B(i)$ *and* $C(i)$;
    $i \leftarrow i+1$;
  end
end

Instead of giving a detailed survey on GAs, we will introduce here an effective GA as the general solution method for finding the optimal solutions of the complex probabilistic portfolio selection problems.

Representation Structure: Since the solution of x_i are required to be $0 \le x_i \le 1$, for $i = 1, 2, \cdots, n$, the solution $\boldsymbol{x} = (x_1, x_2, \cdots, x_n)$ is represented by the chromosome $C = (c_1, c_2, \cdots, c_n)$, where the genes $c_1, c_2, \cdots, c_n$ are restricted in the interval $[0, 1]$. Since it is required that $x_1 + x_2 + \cdots + x_n = 1$, a solution is matched with a chromosome in the following way,

$$x_i = \frac{c_i}{c_1 + c_2 + \cdots + c_n}, \quad i = 1, 2, \cdots, n \tag{2.35}$$

which ensures that $x_1 + x_2 + \cdots + x_n = 1$ always holds.

Initialization: In this procedure, we randomly produce chromosomes from the interval $[0, 1]$ and check their feasibility by stochastic simulation. Repeat this action until feasible *pop_size* chromosomes are produced. The details are as follows:

Randomly generate a chromosome which is composed of n numbers of genes, i.e., randomly generate points $(c_1, c_2, \cdots, c_n)$ from the hypercube $[0, 1]^n$. Calculate the constraint values and check the feasibility of the chromosome. If it is difficult to calculate the constraint values in traditional ways, use stochastic simulation to calculate them. If the chromosome passes the constraints, it is a feasible chromosome. Otherwise, randomly generate points $(c_1, c_2, \cdots, c_n)$ from the hypercube $[0,1]^n$ again until a feasible chromosome is available. Continue this action until feasible *pop_size* chromosomes are produced.

For example, if we use the genetic algorithm to solve the mean-variance model (2.30) in Section 2.5.2 in page 44, we need to randomly generate points $(c_1, c_2, \cdots, c_n)$ from the hypercube $[0,1]^n$. Then calculate the expected value of the chromosome $C = (c_1, c_2, \cdots, c_n)$ (in this example, we can calculate the expected value without simulation). After that, check the feasibility of the chromosome $C = (c_1, c_2, \cdots, c_n)$ as follows:

If $E[x_1\xi_1 + x_2\xi_2 + x_3\xi_3 + x_4\xi_4 + x_5\xi_5 + x_6\xi_6] < 0.1$ *return* 0;
return 1;
in which 1 means feasible, and 0 non-feasible.

Evaluation Function: Evaluation function, denoted by $Eval(C)$, is to assign a probability of reproduction to each chromosome C such that its likelihood of being selected to produce offspring is proportional to its fitness.

There are several kinds of evaluation functions. *Rank-based evaluation function* is one of the most popular ones and is adopted in the algorithm. In the rank based method, the *pop_size* chromosomes are rearranged according to their objective values to make better chromosomes take smaller ordinal numbers. That is, after rearrangement, among *pop_size* chromosomes $C_1, C_2, \cdots, C_{pop\_size}$, C_1 is the best chromosome, and C_{pop_size} the worst one. When calculating the objective value for each chromosome, use stochastic simulation method if it is difficult to calculate the objective value in traditional ways. For example, in the mean-variance model (2.30) in Section 2.5.2 in page 44, we first calculate the objective values, i.e., variance values $V[x_1\xi_1 + x_2\xi_2 + \cdots + x_6\xi_6]$ for all the chromosomes (in this example we can calculate the variance value according to the definition of variance directly without using simulation). Then, for any two chromosomes, the one with lower variance value is the better one and is assigned the smaller ordinal number. The chromosome with the lowest variance value is assigned order one, and the chromosome with the highest variance value is assigned order *pop_size*.

Now given a parameter $a \in (0,1)$ in the genetic system, the *rank-based evaluation function* used in the algorithm is defined as follows,

$$Eval(C_i) = a(1-a)^{i-1}, \qquad i = 1, 2, \cdots, pop\_size. \tag{2.36}$$

Note that $i = 1$ means the best individual, and $i = pop\_size$ means the worst one.

Selection Process: The selection of chromosomes is done by *spinning the roulette wheel* such that the better chromosomes will have more chance to produce offsprings. The selection process is as follows:

First, Compute the cumulative probability p_i for each chromosome C_i,

$$p_0 = 0, \quad p_i = \sum_{k=1}^{i} Eval(C_k), \quad k = 1, \cdots, pop\_size.$$

Here, if we want, we can divide all p_k's, $k = 1, 2, \cdots, pop\_size$, by p_{pop_size} such that $p_{pop\_size} = 1$. Then, randomly generate a real number m from the interval $(0, p_{pop\_size}]$, and the probability of the number m falling in $(p_{k-1}, p_k]$ is the probability that the k-th chromosome will be selected. The probability is proportional to the fitness of the chromosome. However, if we do not divide all p_k's, $k = 1, 2, \cdots, pop\_size$, by p_{pop_size}, no influence is exerted on the genetic process.

Second, randomly generate a real number b from $(0, p_{pop\_size}]$.

Third, Select the i-th chromosome $C_i (1 \le i \le pop\_size)$ if $p_{i-1} < b \le p_i$.

Fourth, Repeat the second to third steps *pop_size* times, and *pop_size* chromosomes are selected.

Crossover Operation: A parameter P_c of a genetic system as the probability of crossover should be predetermined first. The parents for crossover operation are selected by doing the following process repeatedly *pop_size* times: randomly generate a real number d from the interval $[0, 1]$; if $d < P_c$, we take the chromosomes C_i as parents, denoting them by $C'_1, C'_2, C'_3, \cdots$, and dividing them into the following pairs: $(C'_1, C'_2), (C'_3, C'_4), (C'_5, C'_6), \cdots$. Crossover operation on each pair is illustrated through the crossover operation on the pair (C'_1, C'_2). First, we generate a random number e from the open interval $(0, 1)$. Then we produce two new chromosomes X and Y through crossover operator by $X = e \cdot C'_1 + (1 - e) \cdot C'_2, Y = (1 - e) \cdot C'_1 + e \cdot C'_2$. If X and Y are checked to be feasible, we take them as children and replace their parents with them; otherwise, we keep the feasible one if it exists, and then redo the crossover operator by regenerating a random number e until two feasible children are obtained or a given number of cycles is finished. In this case, we only replace the parents with the feasible children.

Mutation Operation: A parameter P_m of a genetic system as the probability of mutation should be predetermined first. In a similar manner as crossover operation, we repeat the following process *pop_size* times to select parents for mutation: randomly generate a real number h from the interval $[0, 1]$; if $h < P_m$, we take the chromosomes C_i as parents for mutation.

Mutation operation on each selected parent is illustrated through the mutation operation on the parent denoted by $C = (c_1, c_2, \cdots, c_n)$. Randomly choose a mutation direction D in $\Re^n$. Let M be an appropriately large positive number. If $C + M \cdot D$ is feasible, we take the new chromosome as the child. Otherwise, we set M as a random number between 0 and M until the new chromosome is feasible. If a feasible solution can not be found in a predetermined number of iterations, we set $M = 0$. Anyway, we replace the parent C with its child $X = C + M \cdot D$.

2.6.4 Hybrid Intelligent Algorithm

After selection, crossover and mutation, the new population is ready for its next evaluation. The hybrid intelligent algorithm will continue until a given number of cyclic repetitions of the above steps is met. We summarize the algorithm as follows:

Step 1. Initialize feasible *pop_size* chromosomes, in which stochastic simulation is used to constraint checking if it is necessary.
Step 2. Calculate the objective values for all chromosomes. Use stochastic simulation if it is necessary.
Step 3. Give the rank order of the chromosomes according to the objective values to make better chromosomes take smaller ordinal numbers, and compute the values of the rank-based evaluation function for all the chromosomes.
Step 4. Compute the fitness of each chromosome according to the rank-based-evaluation function.
Step 5. Select the chromosomes by spinning the roulette wheel.
Step 6. Update the chromosomes by crossover and mutation operations, in which stochastic simulation is used to constraint checking if it is necessary.
Step 7. Repeat the second to the sixth steps for a given number of cycles.
Step 8. Take the best chromosome as the solution of the portfolio selection problem.

2.6.5 Application Example

In the following example, the parameters in the GA are as follows: the population size is 30, the parameter a in the rank-based-evaluation function is 0.05, the probability of mutation is 0.2 and the probability of crossover is 0.3. GAs are very robust in these parameter settings.

Example 2.21. Suppose the investors adopt mean-risk selection idea and want to select portfolio from eight securities. Among them, six security returns are believed to be normally distributed random variables. The rest two securities are newly listed securities. The investors are not sure what kinds of variable the two security returns are but believe that their returns fall in a certain interval. Therefore, they use uniformly distributed random variables to describe them. The eight security returns are given in Table 2.15. Since the portfolio return containing these eight securities is neither a normal random variable nor a uniform random variable, it is difficult to use "Excel" to solve the mean-risk selection problem. In this case, we can use the preceding introduced hybrid intelligent algorithm to find the optimal portfolio.

Table 2.15 Random Returns of Eight Securities

Security i	Random Return ξ_i	Security i	Random Return ξ_i
1	$\mathcal{N}(0.1042, 0.0567)$	5	$\mathcal{N}(0.1093, 0.0582)$
2	$\mathcal{N}(0.0878, 0.0590)$	6	$\mathcal{N}(0.1064, 0.0648)$
3	$\mathcal{N}(0.0754, 0.0112)$	7	$\mathcal{U}(-0.4, 0.6)$
4	$\mathcal{N}(0.0809, 0.0211)$	8	$\mathcal{U}(-0.35, 0.53)$

Suppose the risk-free interest rate $r_f = 0.003$, and the investors' confidence curve is

$$\alpha(r) = \begin{cases} -1.5r + 0.28, & \text{when } 0 \le r \le 0.12 \\ -0.5r + 0.16, & \text{when } 0.12 \le r \le 0.3 \\ 0.01, & \text{when } r \ge 0.3. \end{cases}$$

Then the mean-risk selection model is as follows:

$$\begin{cases} \max E[\xi_1 x_1 + \xi_2 x_2 + \xi_3 x_3 + \xi_4 x_4 + \xi_5 x_5 + \xi_6 x_6 + \xi_7 x_7 + \xi_8 x_8] \\ \text{subject to:} \\ \quad R(x_1, x_2, \cdots, x_8; r) \le \alpha(r), \ \forall r \ge 0 \\ \quad x_1 + x_2 + x_3 + x_4 + x_5 + x_6 + x_7 + x_8 = 1 \\ \quad x_1, x_2, x_3, x_4, x_5, x_6, x_7, x_8 \ge 0 \end{cases} \tag{2.37}$$

where

$$R(x_1, x_2, \cdots, x_8; r) = \Pr\{0.003 - (\xi_1 x_1 + \xi_2 x_2 + \cdots + \xi_7 x_7 + \xi_8 x_8) \ge r\}.$$

Since the confidence curve is a horizontal line when $r \ge 0.3$, and the risk curve is a decreasing function of r, we just need to check if the risk curve of the portfolio is below the investors' confidence curve for $r \in [0, 0.3]$. As an approximation, we set $r = 0, r = 0.01, r = 0.02, r = 0.03, \cdots, r = 0.3$. That is, to find the optimal portfolio, we solve the following model

$$\begin{cases} \max E[\xi_1 x_1 + \xi_2 x_2 + \cdots + \xi_8 x_8] \\ \text{subject to:} \\ \quad \Pr\{0.003 - (\xi_1 x_1 + \xi_2 x_2 + \cdots + \xi_8 x_8) \ge 0\} \le 0.28 \\ \quad \Pr\{0.003 - (\xi_1 x_1 + \xi_2 x_2 + \cdots + \xi_8 x_8) \ge 0.01\} \le 0.265 \\ \quad \Pr\{0.003 - (\xi_1 x_1 + \xi_2 x_2 + \cdots + \xi_8 x_8) \ge 0.02\} \le 25 \\ \quad \cdots \\ \quad \Pr\{0.003 - (\xi_1 x_1 + \xi_2 x_2 + \cdots + \xi_8 x_8) \ge 0.3\} \le 0.01 \\ \quad x_1 + x_2 + \cdots + x_8 = 1 \\ \quad x_1, x_2, \cdots, x_8 \ge 0. \end{cases} \tag{2.38}$$

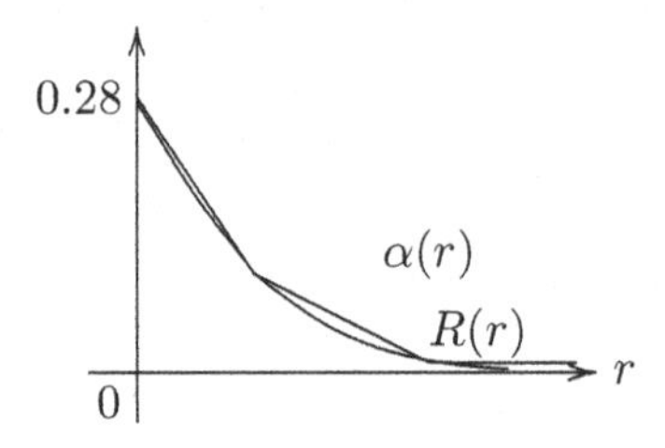

Fig. 2.19 Risk curve $R(r)$ and confidence curve $\alpha(r)$ of Example 2.21.

For $r_i = 0.01i, i = 0, 1, 2, \cdots, 30$, to check if

$$\Pr\{0.003 - (\xi_1 x_1 + \xi_2 x_2 + \cdots + \xi_8 x_8) \geq r_i\} \leq \alpha(r_i),$$

first, generate random numbers from normal and uniform distributions of the corresponding security returns and use stochastic simulation to calculate the probability value of

$$\Pr\{0.003 - (\xi_1 x_1 + \xi_2 x_2 + \cdots + \xi_8 x_8) \geq r_i\},\ r_i = 0.01i,\ i = 0, 1, 2, \cdots, 30.$$

Next, integrate the simulation results into the GA to produce the hybrid intelligent algorithm. A run of the algorithm (3,000 cycles in simulation, 4,000 generations in the GA) shows that the investors should allocate their money according to Table 2.16. The maximum expected value is 10.83%. The risk curve and confidence curve are drawn in Fig. 2.19. It can be seen that the risk curve of the portfolio is below the investors' confidence curve. The hybrid intelligent algorithm is summarized as follows:

Table 2.16 Allocation of Money to Eight Securities (%)

x_1	x_2	x_3	x_4	x_5	x_6	x_7	x_8
3.09	0.00	0.00	0.00	68.19	28.72	0.00	0.00

Hybrid Intelligent Algorithm

Step 1. Determine representation structure of solutions by chromosomes. Since the solution of x_i are required to be $0 \leq x_i \leq 1$, for $i = 1, 2, \cdots, n$, the solution $\boldsymbol{x} = (x_1, x_2, \cdots, x_n)$ is represented by the chromosome $C = (c_1, c_2, \cdots, c_n)$, where the genes $c_1, c_2, \cdots, c_n$ are restricted in the interval $[0, 1]$. Since it is required that $x_1 + x_2 + \cdots + x_n = 1$, a solution is matched with a chromosome in the following way,

$$x_i = \frac{c_i}{c_1 + c_2 + \cdots + c_n}, \quad i = 1, 2, \cdots, n$$

which ensures that $x_1 + x_2 + \cdots + x_n = 1$ always holds.

Step 2. Set parameters $P_c = 0.3, P_m = 0.2, pop\_size = 30$ in the GA.
Step 3. Generate the chromosomes $C = (c_1, c_2, \cdots, c_8)$ from $[0, 1]^8$.
Step 4. Use stochastic simulation to calculate the point values on the risk curve of the portfolio for each chromosome. The simulation procedures for calculating $\Pr\{0.003 - (\xi_1 x_1 + \xi_2 x_2 + \cdots + \xi_8 x_8) \geq r_i\}, r_i = 0.01i, i = 0, 1, \cdots, 30$, are similar to Example 2.16 in page 50. Then check the feasibility of the chromosome as follows:

If $\Pr\{0.003 - (\xi_1 x_1 + \xi_2 x_2 + \cdots + \xi_8 x_8) \geq 0\} \leq 0.28$

$\Pr\{0.003 - (\xi_1 x_1 + \xi_2 x_2 + \cdots + \xi_8 x_8) \geq 0.01\} \leq 0.265$

$\Pr\{0.003 - (\xi_1 x_1 + \xi_2 x_2 + \cdots + \xi_8 x_8) \geq 0.02\} \leq 0.25$

...

$\Pr\{0.003 - (\xi_1 x_1 + \xi_2 x_2 + \cdots + \xi_8 x_8) \geq 0.3\} \leq 0.01$

return 1;
return 0;
in which 1 means feasible, and 0 non-feasible.

Step 5. Repeat the third and fourth steps until feasible *pop_size* numbers of chromosomes are produced.
Step 6. Calculate the objective values and give the rank order of the chromosomes according to the objective values to make better chromosomes take smaller ordinal numbers.
Step 7. Compute the values of the rank-based evaluation function for all the chromosomes.
Step 8. Calculate the fitness of each chromosome according to the rank-based-evaluation function.
Step 9. Select the chromosomes by spinning the roulette wheel.
Step 10. Update the chromosomes by crossover and mutation operations.
Step 11. Repeat the sixth to the tenth steps for 4000 cycles.
Step 12. Take the best chromosome as the solution of the portfolio selection problem.

2.7 Remarks

This chapter has introduced four basic types of portfolio selection models. They are mean-risk model, β-return-risk model, probability minimization model, mean-variance model and its improvement mean-semivariance model. Different models reflect the different decision attitudes of the investors and the selection of the models depends on the investors' preference and their information processing ability. Generally speaking, the investors who adopt risk curve are the most cautious investors. They evaluate each likely loss event and compare it with their own tolerance ability. Therefore, the decision making based on risk curve is the safest. In addition, risk curve provides instinct loss levels. With mean-risk model, the investors require a maximum average return; with β-return-risk model, the investors want a maximum specific target return value.

Probability minimization model can be regarded as a degeneration version of the mean-risk or β-return-risk model because in the model, the investors only concern and deal with one sensitive loss event instead of all the likely loss events. Without needing to find out the investors' confidence curve and probability values for big enough numbers of loss events, information processing work is much less. With probability minimization model, the investors may lose far more than the specific loss amount, but they won't do so very often.

Mean-variance model, on the other hand, uses average information to evaluate risk. Using mean-variance model, the investors do not need to know the probability distribution of the portfolio return. With the expected returns and variance values of individual security returns, the investors can easily obtain the expected value and variance value of any portfolios. However, using mean-variance model, the investors have to tell their tolerable maximum variance level or minimum expected value or give their utility function. In addition, when using mean-variance model, the investors have to make sure that the portfolio return is symmetrically distributed, while using other models, the symmetry requirement is not a necessity.

Chapter 3
Credibilistic Portfolio Selection

Credibilistic portfolio selection deals with fuzzy portfolio selection by means of credibility theory. Fuzzy portfolio selection problem was researched from 1990s. Early researchers employed possibility as the basic measure of the occurrence of a fuzzy event and most of them devoted themselves to extending Markowitz's mean-variance selection idea. However, possibility measure is not self-dual. By using possibility, when the investors know the possibility level of a portfolio reaching a target return, they cannot know the possibility level of the opposite event, i.e., the event of this portfolio not being able to achieve the target return! This will confuse and worry the decision maker. Therefore, Huang proposed that we should use the self-dual credibility as the basic measure of the occurrence of a fuzzy event and study the fuzzy portfolio selection problems. To provide an instinct and observable information about loss amount and to accurately evaluate the loss degree, Huang [38] proposed that we should evaluate each likely loss level and the loss occurrence chance instead of just focusing on the average information of loss. Looking at loss from a panoramic perspective, Huang provided a general definition of risk, i.e., the risk curve, and proposed a mean-risk model based on this new definition. In addition, Huang proposed a spectrum of simplified versions of the risk and proposed a system of credibilistic portfolio selection models [41] including mean-risk model [38], β-return-risk model [27], credibility minimization model, mean-variance model [33], mean-semivariance model [37], and entropy optimization model [39].

This chapter will first introduce some necessary knowledge about credibility theory. The reason for adopting credibility measure rather than possibility measure is given. Then we will introduce the definitions of risk and the credibilistic portfolio selection models. Crisp equivalents of the fuzzy models in some special cases will also be presented. After that, we will provide a general solution algorithm for solving the credibilistic portfolio selection models.

X. Huang: Portfolio Analysis: From Probab. to Credibilistic, STUDFUZZ 250, pp. 61–115.
springerlink.com

3.1 Fundamentals of Credibility Theory

In reality, besides randomness, there are many fuzzy phenomena. For example, a "beautiful" girl in many situation is not a very clear concept. In order to describe fuzziness, the concept of fuzzy set was first proposed by Zadeh [95] in 1965 via membership function. Furthermore, to measure a fuzzy event, Zadeh [96] proposed possibility measure. Although possibility is a widely used measure, it is not self-dual. However, for a measure, self-duality property is extremely important. In order to define a self-dual measure, Liu and Liu [55] proposed credibility measure. An axiomatic credibility theory was founded by Liu [58] in 2004 and refined by Liu [60] in 2007. Credibility theory has been fairly well applied in many application areas.

Credibility and Credibility Space

Definition 3.1 *(Liu [60]). Let Θ be a nonempty set, and $\mathcal{P}(\Theta)$ the power set of Θ, i.e., the largest σ-algebra over Θ. Each element in $\mathcal{P}(\Theta)$ is called an event. The set function* Cr *is called a credibility measure if*
(Axiom 1) (Normality) $\mathrm{Cr}\{\Theta\} = 1$;
(Axiom 2) (Monotonicity) $\mathrm{Cr}\{A\} \leq \mathrm{Cr}\{B\}$ *whenever* $A \subset B$;
(Axiom 3) (Self-duality) $\mathrm{Cr}\{A\} + \mathrm{Cr}\{A^c\} = 1$ *for any event* A.
(Axiom 4) (Maximality) $\mathrm{Cr}\{\cup_i A_i\} = \sup_i \mathrm{Cr}\{A_i\}$ *for any events* $\{A_i\}$ *with* $\sup_i \mathrm{Cr}\{A_i\} < 0.5$.

The value of $\mathrm{Cr}\{A\}$ indicates the level that the event A will occur.

For example, let $\Theta = \{\theta_1, \theta_2\}$. There are only four events: $\emptyset, \{\theta_1\}, \{\theta_2\}, \Theta$. Define $\mathrm{Cr}\{\emptyset\} = 0, \mathrm{Cr}\{\theta_1\} = 0.4, \mathrm{Cr}\{\theta_2\} = 0.6$, and $\mathrm{Cr}\{\Theta\} = 1$. Then the set function Cr is a credibility measure because it satisfies the four axioms.

Let Θ be a nonempty set, $\mathcal{P}$ the power set of Θ, and Cr the credibility measure. From Axioms 1 and 3 we know $\mathrm{Cr}\{\emptyset\} = 0$. From Axiom 2 we know $0 \leq \mathrm{Cr}\{A\} \leq 1$ for any $A \in \mathcal{P}$ because $\emptyset \subset A \subset \Theta$. That is, the credibility value of a fuzzy event is in the interval $[0, 1]$.

Definition 3.2 *(Liu [60]). Let Θ be a nonempty set, $\mathcal{P}(\Theta)$ the power set of Θ, and* Cr *a credibility measure. Then the triplet* $(\Theta, \mathcal{P}(\Theta), \mathrm{Cr})$ *is called a credibility space.*

Fuzzy Variable

Definition 3.3. *A fuzzy variable is defined as a function from a credibility space* $(\Theta, \mathcal{P}(\Theta), \mathrm{Cr})$ *to the set of real numbers.*

Remark 3.1. Since $\mathcal{P}(\Theta)$ is the power set of Θ (i.e., the collection of all the subsets of Θ), and a fuzzy variable ξ is a function on a credibility space, for any set B of real numbers, the set

$$\{\xi \in B\} = \{\theta \in \Theta | \xi(\theta) \in B\}$$

is always an element in $\mathcal{P}$. That is, the fuzzy variable is a measurable function and $\{\xi \in B\}$ is an event.

Definition 3.4. *Let ξ_1 and ξ_2 be two fuzzy variables defined on the credibility space $(\Theta, \mathcal{P}, \mathrm{Cr})$. We say $\xi_1 = \xi_2$ if $\xi_1(\theta) = \xi_2(\theta)$ for almost all $\theta \in \Theta$.*

Membership Function and Credibility Inversion Theorem

The membership function was first introduced by Zadeh [95] in 1965. In the credibility theory, membership function was defined via credibility.

Definition 3.5 *(Liu [60]). Let ξ be a fuzzy variable defined on the credibility space $(\Theta, \mathcal{P}(\Theta), \mathrm{Cr})$. Then its membership function is derived from the credibility measure by*

$$\mu(t) = (2\mathrm{Cr}\{\xi = t\}) \wedge 1, \quad t \in \Re.$$

If we have got the membership function of a fuzzy variable ξ first, how can we know the credibility degree of a fuzzy event? The following inversion theorem gives the answer.

Theorem 3.1 *(Credibility Inversion Theorem, Liu and Liu [55]). Let ξ be a fuzzy variable with membership function μ. Then for any set A of real numbers, we have*

$$\mathrm{Cr}\{\xi \in A\} = \frac{1}{2}\left(\sup_{t \in A} \mu(t) + 1 - \sup_{t \in A^c} \mu(t)\right). \tag{3.1}$$

Proof: If $\mathrm{Cr}\{\xi \in A\} \le 0.5$, we know from Axiom 2 that $\mathrm{Cr}\{\xi = t\} \le 0.5$ for each $t \in A$. According to Axiom 4 we have

$$\mathrm{Cr}\{\xi \in A\} = \frac{1}{2}(\sup_{t \in A}(2\mathrm{Cr}\{\xi = t\} \wedge 1)) = \frac{1}{2}\sup_{t \in A} \mu(t). \tag{3.2}$$

Since the credibility measure is self-dual, we have $\mathrm{Cr}\{\xi \in A^c\} \ge 0.5$, and $\sup_{t \in A^c} \mathrm{Cr}\{\xi = t\} \ge 0.5$. Therefore

$$\sup_{t \in A^c} \mu(t) = \sup_{t \in A^c}(2\mathrm{Cr}\{\xi = t\} \wedge 1) = 1. \tag{3.3}$$

It follows from (3.2) and (3.3) that (3.1) holds.

If $\mathrm{Cr}\{\xi \in A\} \geq 0.5$, we have $\mathrm{Cr}\{\xi \in A^c\} \leq 0.5$ because the credibility measure is self-dual. From the result of the first case we have

$$\mathrm{Cr}\{\xi \in A\} = 1 - \mathrm{Cr}\{\xi \in A^c\} = 1 - \frac{1}{2}\left(\sup_{t \in A^c} \mu(t) + 1 - \sup_{t \in A} \mu(t)\right)$$

$$= \frac{1}{2}\left(\sup_{t \in A} \mu(t) + 1 - \sup_{t \in A^c} \mu(t)\right).$$

The theorem is proven.

Example 3.1. Let ξ be a fuzzy variable with membership function μ. Then it follows from Theorem 3.1 that the following equations hold:

$$\mathrm{Cr}\{\xi = t\} = \frac{1}{2}\left(\mu(t) + 1 - \sup_{y \neq t} \mu(y)\right), \quad \forall t \in \Re; \tag{3.4}$$

$$\mathrm{Cr}\{\xi \leq t\} = \frac{1}{2}\left(\sup_{y \leq t} \mu(y) + 1 - \sup_{y > t} \mu(y)\right), \quad \forall t \in \Re; \tag{3.5}$$

$$\mathrm{Cr}\{\xi \geq t\} = \frac{1}{2}\left(\sup_{y \geq t} \mu(y) + 1 - \sup_{y < t} \mu(y)\right), \quad \forall t \in \Re. \tag{3.6}$$

Especially, if μ is a continuous function, we have

$$\mathrm{Cr}\{\xi = t\} = \frac{\mu(t)}{2}, \quad \forall t \in \Re. \tag{3.7}$$

Remark 3.2. A fuzzy variable has a unique membership function, but a membership function may produce multiple fuzzy variables. For example, let $\Theta = \{\theta_1, \theta_2\}$ and $\mu(\theta_1) = \mu(\theta_2) = 1$. It can be easily proven that $(\Theta, \mathcal{P}, \mathrm{Cr})$ is a credibility space. Define

$$\xi_1(\theta) = \begin{cases} 0, & \text{if } \theta = \theta_1 \\ 1, & \text{if } \theta = \theta_2, \end{cases} \qquad \xi_2(\theta) = \begin{cases} 1, & \text{if } \theta = \theta_1 \\ 0, & \text{if } \theta = \theta_2. \end{cases}$$

We can see that though the fuzzy variables ξ_1 and ξ_2 have the same membership function, i.e., $\mu(t) \equiv 1$ on $t = 0$ or 1, they are two different fuzzy variables in the sense of Definition 3.4. Since one membership function may produce multiple fuzzy variables, we can not define fuzzy variable via membership function. An axiomatic system is needed to define a fuzzy variable and discuss the properties concerning the fuzzy variable to ensure precision and consistency of the researches. That explains why the membership function is defined via credibility in credibility theory. However, for application purposes, we only need to construct the membership function of a fuzzy variable and then use the credibility inversion theorem to derive the credibility and use credibility theory to help solve the application problems. The mathematical requirement for the membership function is simple. It has been proven [58] that a function $\mu :\rightarrow [0, 1]$ is a membership function if and only if

$\sup \mu(t) = 1$. For construction method of membership functions, readers can refer to Triantaphyllou and Mann [90], Chen and Otto [8], Kumar and Ganesh [47], Hong and Chen [26], and Medaglia, Fang, Nuttle and Wilson [70].

Remark 3.3. Membership function indicates the degree that the fuzzy variable ξ takes some prescribed values. If t is an impossible point, the membership degree $\mu(t) = 0$; and if t is the most possible point that the fuzzy variable ξ takes, the membership degree $\mu(t) = 1$. However, the inverse statement is not true. It is the credibility degree rather than membership degree that gives the occurrence chance of the prescribed values. From credibility inversion theorem we know that the credibility degree of a prescribed value depends not only on its membership degree but also on the membership degree of its complementary set.

Why Adopt Credibility?

Possibility measure is an early proposed measure to measure a fuzzy event. Let ξ be a fuzzy variable with membership function μ. Then $\text{Pos}\{A\} = \sup\{\mu(\xi(\theta))|\theta \in A\}$ for any fuzzy event $A \in \mathcal{P}$. Though possibility measure is an important measure and is widely used in fuzzy set theory, it is not self-dual. Yet, self-duality property is absolutely needed in both theory and application research. Without self-duality, confusion will appear. Let us see below what will happen if we adopt possibility to measure the occurrence chance of a fuzzy event.

Example 3.2. A fuzzy variable is called a *triangular fuzzy variable* if it has a triangular membership function (see Fig. 3.1)

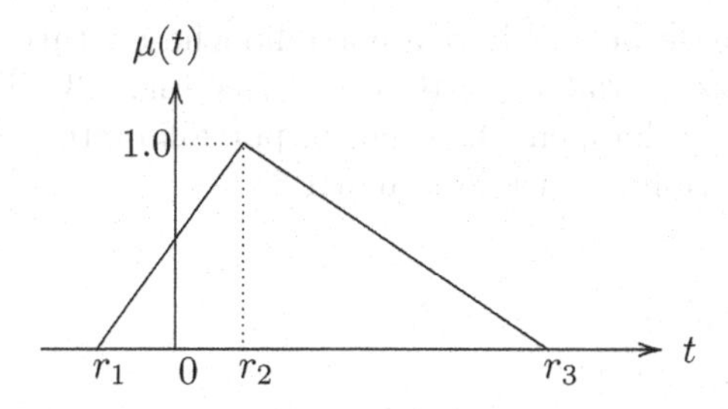

Fig. 3.1 Triangular membership function.

$$\mu(t) = \begin{cases} \dfrac{t - r_1}{r_2 - r_1}, & \text{if } r_1 \le t \le r_2 \\ \dfrac{t - r_3}{r_2 - r_3}, & \text{if } r_2 \le t \le r_3 \\ 0, & \text{otherwise.} \end{cases}$$

We denote it by $\xi = (r_1, r_2, r_3)$ with $r_1 < r_2 < r_3$.

Question 1: Suppose a traveler is going to visit a city. The expenditure is predicted to be a triangular fuzzy variable $\xi = (200, 300, 400)$ dollars (see Fig. 3.2). To ensure that the traveler will have enough money in his traveling, at least how much money should he bring with him?

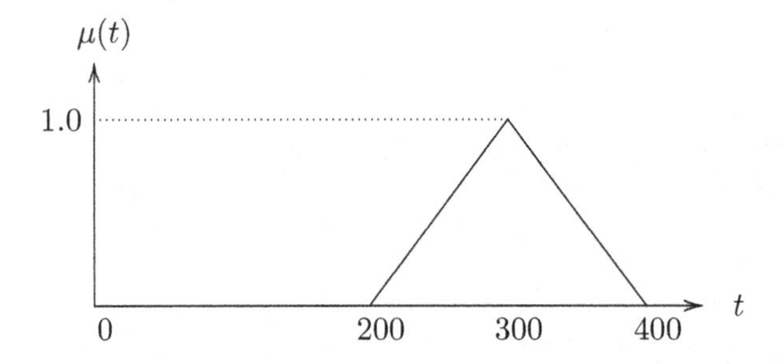

Fig. 3.2 Triangular expenditure $\xi = (200, 300, 400)$.

Judging from common sense, we will say that the traveler should bring 400 dollars with him so that he will have enough money for traveling. However, when using possibility measure, we find

$$\min\left\{t|\text{Pos}\{\xi \le t\} = 1\right\} = 300,$$

which tells us that the traveler just needs to bring with him 300 dollars to ensure that he will have enough money for traveling. This result obviously is contradictory to our judgement and common sense.

Question 2: Suppose we now have a portfolio whose return can be described by a triangular fuzzy variable $\xi = (0, 1.5, 3)$ (see Fig. 3.3). Then which event will be more likely to happen, the event of portfolio return not less than 1.5 or the event of portfolio return less than 1.5?

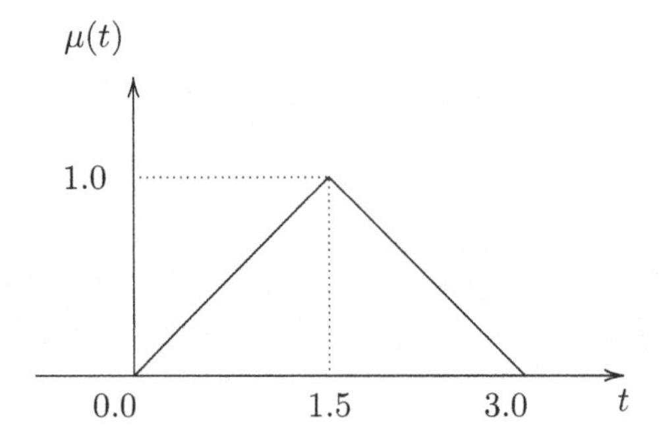

Fig. 3.3 Triangular portfolio return $\xi = (0, 1.5, 3)$.

By using possibility measure, we can calculate that $\text{Pos}\{\xi \geq 1.5\} = 1$, which seems to imply that the portfolio return not less than the value 1.5 will surely happen. However, by using possibility measure, we can also calculate that $\text{Pos}\{\xi < 1.5\} = 1$, which seems to imply that the portfolio return less than the value 1.5 will also surely happen. Is not it strange that two opposite events will both surely happen at the same time? The law of contradiction tells us that a proposition cannot be both true and false at the same time, and the law of excluded middle says that a proposition should be either true or false. It is obviously that the judgement made based on possibility is in contradiction with both the law of contradiction and the law of excluded middle.

Now, let us use credibility measure to calculate a fuzzy event. It follows from credibility inversion theorem that for a triangular fuzzy variable $\xi = (r_1, r_2, r_3)$ (Fig. 3.1), we have

$$\text{Cr}\{\xi \leq t\} = \begin{cases} 1, & r_3 \leq t \\ \dfrac{r_3 - 2r_2 + t}{2(r_3 - r_2)}, & r_2 \leq t \leq r_3 \\ \dfrac{t - r_1}{2(r_2 - r_1)}, & r_1 \leq t \leq r_2 \\ 0, & \text{otherwise.} \end{cases} \tag{3.8}$$

Then, for a triangular expenditure $\xi = (200, 300, 400)$, we have

$$\min\left\{t \middle| \text{Cr}\{\xi \leq t\} = 1\right\} = 400,$$

which means that the traveler should bring with him 400 dollars to ensure that he will have enough money for traveling. The result is consistent with our judgement and common sense.

For a triangular portfolio return $\xi = (0, 1.5, 3)$, according to Equation (3.8), we have $\text{Cr}\{\xi \geq 1.5\} = 0.5$, which means that there is only half the chance that the portfolio return will not be less than 1.5. According to Equation (3.8), we know $\text{Cr}\{\xi < 1.5\} = 0.5$, which means that there is only half the chance that the portfolio return will be less than 1.5. It is seen that the result is consistent with our judgement and the confusion disappears.

Some Special Fuzzy Variables

Example 3.3. A fuzzy variable is called a *trapezoidal fuzzy variable* if it has a trapezoidal membership function (see Fig. 3.4)

$$\mu(t) = \begin{cases} \dfrac{t - r_1}{r_2 - r_1}, & \text{if } r_1 \le t \le r_2 \\ 1, & \text{if } r_2 \le t \le r_3 \\ \dfrac{t - r_4}{r_3 - r_4}, & \text{if } r_3 \le t \le r_4 \\ 0, & \text{otherwise.} \end{cases}$$

We denote it by $\xi = (r_1, r_2, r_3, r_4)$ with $r_1 < r_2 < r_3 < r_4$.

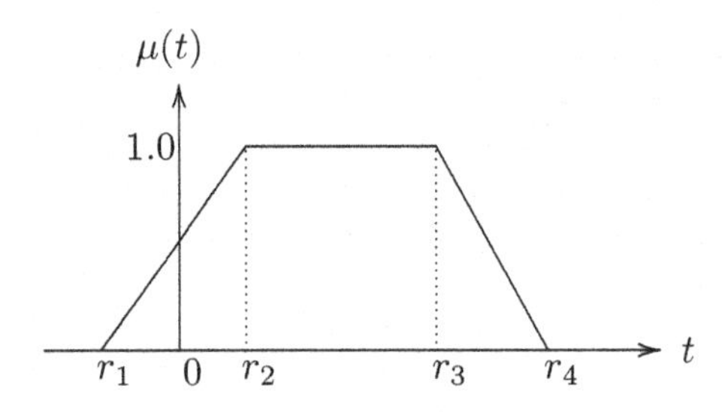

Fig. 3.4 Trapezoidal membership function.

According to credibility inversion theorem, if $r_4 \le t$, we have

$$\mathrm{Cr}\{\xi \le t\} = \frac{1}{2}(1 + 1 - 0) = 1.$$

If $r_3 \le t \le r_4$, then

$$\mathrm{Cr}\{\xi \le t\} = \frac{1}{2}(1 + 1 - \frac{r_4 - t}{r_4 - r_3}) = \frac{r_4 - 2r_3 + t}{2(r_4 - r_3)}.$$

If $r_2 \le t \le r_3$, then

$$\mathrm{Cr}\{\xi \le t\} = \frac{1}{2}(1 + 1 - 1) = \frac{1}{2}.$$

If $r_1 \le t \le r_2$, then

$$\mathrm{Cr}\{\xi \le t\} = \frac{1}{2}(\frac{t - r_1}{r_2 - r_1} + 1 - 1) = \frac{t - r_1}{2(r_2 - r_1)}.$$

If $t < r_1$, then

$$\mathrm{Cr}\{\xi \le t\} = \frac{1}{2}(0 + 1 - 1) = 0.$$

That is,

$$\operatorname{Cr}\{\xi \le t\} = \begin{cases} 1, & \text{if } r_4 \le t \\ \dfrac{r_4 - 2r_3 + t}{2(r_4 - r_3)}, & \text{if } r_3 \le t \le r_4 \\ \dfrac{1}{2}, & \text{if } r_2 \le t \le r_3 \\ \dfrac{t - r_1}{2(r_2 - r_1)}, & \text{if } x_1 \le t \le r_2 \\ 0, & \text{otherwise.} \end{cases} \tag{3.9}$$

Example 3.4. A fuzzy variable ξ is called a *normal fuzzy variable* if it has a normal membership function

$$\mu(t) = 2\left(1 + \exp\left(\frac{\pi|t-e|}{\sqrt{6}\sigma}\right)\right)^{-1}, \quad t \in R, \quad \sigma > 0.$$

We denote it by $\xi \sim \mathcal{N}(e, \sigma)$. It can be calculated that

$$\mu(e+\sigma) = \mu(e-\sigma) = 0.4324, \quad \text{and} \quad \mu(e+2\sigma) = \mu(e-2\sigma) = 0.1428.$$

Two normal membership functions with same σ but different e's are drawn in Fig. 3.5, and two normal membership functions with same e but different σ's are drawn in Fig. 3.6.

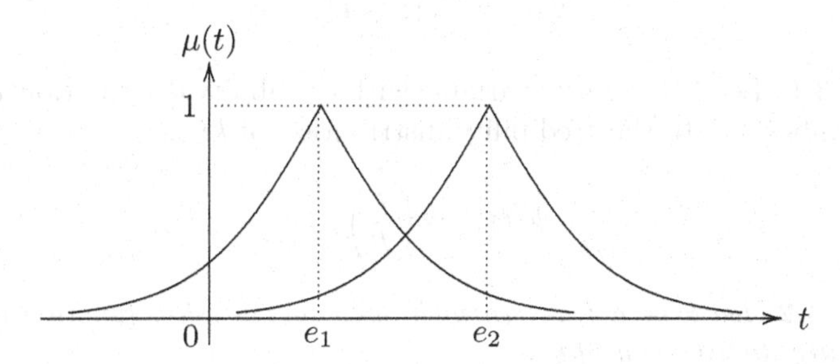

Fig. 3.5 Normal membership functions with same σ but different e's.

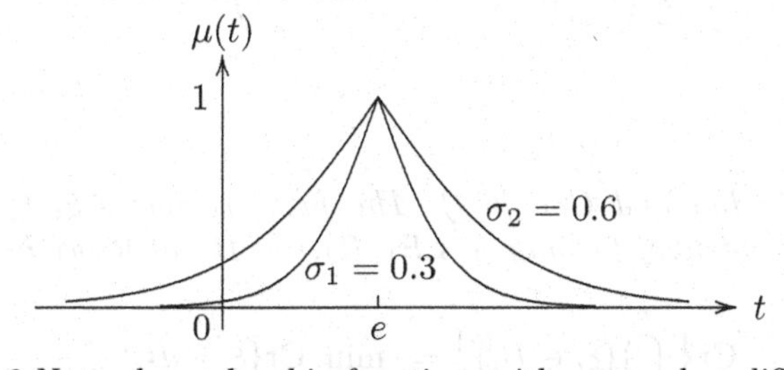

Fig. 3.6 Normal membership functions with same e but different σ's.

Example 3.5. A fuzzy variable ξ is called an *equipossible fuzzy variable* on $[a,b]$ if it has the following membership function (see Fig. 3.7)

$$\mu(t) = \begin{cases} 1, & \text{if } a_1 \le t \le a_2 \\ 0, & \text{otherwise.} \end{cases}$$

We denote it by $\xi = (a,b)$.

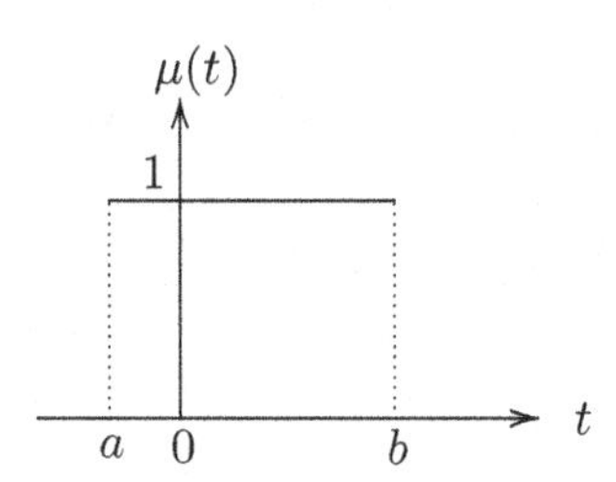

Fig. 3.7 Membership function of an equipossible fuzzy variable.

Credibility Distribution

Definition 3.6 *(Liu [56]). The credibility distribution $\Phi : \Re \to [0,1]$ of a fuzzy variable ξ is defined by*

$$\Phi(t) = \mathrm{Cr}\{\xi \le t\}. \tag{3.10}$$

Example 3.6. Let ξ be a fuzzy variable with credibility distribution Φ. Then for any number $k > 0$, the credibility distribution of $k\xi$ is

$$\Psi(t) = \Phi\left(\frac{t}{k}\right). \tag{3.11}$$

Theorem 3.2. *Let ξ be a fuzzy variable with membership function μ. Then the credibility distribution of ξ is*

$$\Phi(t) = \frac{1}{2}\left(\sup_{z \le t} \mu(z) + 1 - \sup_{z > t} \mu(z)\right), \quad \forall t \in \Re. \tag{3.12}$$

Independence

Definition 3.7 *(Liu and Gao [61]). The fuzzy variables $\xi_1, \xi_2, \cdots, \xi_n$ are said to be independent if for any sets $B_1, B_2, \cdots, B_n$ of $\Re$, we have*

$$\mathrm{Cr}\Big\{\bigcap_{i=1}^{n}\{\xi_i \in B_i\}\Big\} = \min_{1 \le i \le n} \mathrm{Cr}\{\xi_i \in B_i\}. \tag{3.13}$$

Theorem 3.3 *(Liu [60]). The fuzzy variables $\xi_1, \xi_2, \cdots, \xi_n$ are independent if and only if*

$$\mathrm{Cr}\left\{\bigcup_{i=1}^{n}\{\xi_i \in B_i\}\right\} = \max_{1\le i\le n} \mathrm{Cr}\{\xi_i \in B_i\}. \tag{3.14}$$

Proof: Since credibility measure is self-dual, the fuzzy variables $\xi_1, \xi_2, \cdots, \xi_n$ are independent if and only if

$$\begin{aligned}\mathrm{Cr}\left\{\bigcup_{i=1}^{n}\{\xi_i \in B_i\}\right\} &= 1 - \mathrm{Cr}\left\{\bigcap_{i=1}^{n}\{\xi_i \in B_i^c\}\right\} \\ &= 1 - \min_{1\le i\le n} \mathrm{Cr}\{\xi_i \in B_i^c\} = \max_{1\le i\le n} \mathrm{Cr}\{\xi_i \in B_i\}.\end{aligned}$$

Thus, the theorem is proven.

Fuzzy Arithmetic

Definition 3.8 *(Liu [60]). Let $\Re^n \to \Re$ be a function, and $\xi_1, \xi_2, \cdots, \xi_n$ fuzzy variables defined on the credibility space $(\Theta_i, \mathcal{P}(\Theta_i), \mathrm{Cr}_i), i = 1, 2, \cdots, n$, respectively. Then $\xi = f(\xi_1, \xi_2, \cdots, x_n)$ is a fuzzy variable defined as*

$$\xi(\theta) = f\Big(\xi_1(\theta), \xi_2(\theta_2), \cdots, \xi_n(\theta)\Big)$$

for any $\theta \in \Theta$.

Theorem 3.4 *(Extension Principle of Zadeh). Let $\xi_1, \xi_2, \cdots, \xi_n$ be independent fuzzy variables with membership functions $\mu_1, \mu_2, \cdots, \mu_n$, respectively, and $\Re^n \to \Re$ a continuous function. Then the membership function μ of $\xi = f(\xi_1, \xi_2, \cdots, \xi_n)$ is derived from the membership functions $\mu_1, \mu_2, \cdots, \mu_n$ for any $t \in \Re$ by*

$$\mu(t) = \sup_{t=f(t_1,t_2,\cdots,t_n)} \min_{1\le i\le n} \mu_i(t_i) \tag{3.15}$$

which is consistent with the expression

$$\mu(t) = \sup_{t_1,t_2,\cdots,t_n\in\Re} \left\{\min_{1\le i\le n} \mu_i(t_i) \;\Big|\; t = f(t_1, t_2, \cdots, t_n)\right\}. \tag{3.16}$$

Here we set $\mu(t) = 0$ if there are not real numbers $t_1, t_2, \cdots, t_n$ such that $t = f(t_1, t_2, \cdots, t_n)$.

Let us now give some examples to show the operations on fuzzy variables.

Example 3.7. Let ξ be a fuzzy variable with membership function ν. Then the membership function μ of $\xi + 2$ is

$$\begin{aligned}\mu(t) &= \{\nu(t_1) | t = t_1 + 2\} \\ &= \nu(t - 2).\end{aligned}$$

That is, the membership value that the fuzzy variable $\xi + 2$ achieves value $t \in \Re$ is the membership value that the fuzzy variable ξ achieves value $t - 2$.

Example 3.8. Let ξ_1 be a fuzzy variable with membership function μ_1, and ξ_2 another fuzzy variable with membership function μ_2. Then the membership function μ of $\xi_1 \cdot \xi_2$ is

$$\mu(t) = \sup_{t_1, t_2 \in R} \{\mu_1(t_1) \wedge \mu_2(t_2) | t = t_1 \cdot t_2\}.$$

Example 3.9. Let ξ_1 be a fuzzy variable with membership function μ_1, and ξ_2 another fuzzy variable with membership function μ_2. Then the membership function μ of $\xi_1 + \xi_2$ is

$$\begin{aligned} \mu(t) &= \sup_{t_1, t_2 \in R} \{\mu_1(t_1) \wedge \mu_2(t_2) | t = t_1 + t_2\} \\ &= \sup_{t_1 \in R} \{\mu_1(t_1) \wedge \mu_2(t - t_1)\}. \end{aligned}$$

Example 3.10. Let $\xi_1 = (a_1, a_2, a_3, a_4)$ and $\xi_2 = (b_1, b_2, b_3, b_4)$ be two trapezoidal fuzzy variables with membership functions μ_1 and μ_2 respectively. Then the membership function μ of $\xi_1 + \xi_2$ is

$$\begin{aligned} \mu(t) &= \sup_{t_1, t_2 \in R} \{\mu_1(t_1) \wedge \mu_2(t_2) | t = t_1 + t_2\} \\ &= \begin{cases} \dfrac{t - (a_1 + b_1)}{(a_2 + b_2) - (a_1 + b_1)}, & \text{if } a_1 + b_1 \le t \le a_2 + b_2 \\ 1, & \text{if } a_2 + b_2 \le t \le a_3 + b_3 \\ \dfrac{t - (a_4 + b_4)}{(a_3 + b_3) - (a_4 + b_4)}, & \text{if } a_3 + b_3 \le t \le a_4 + b_4 \\ 0, & \text{otherwise,} \end{cases} \end{aligned}$$

which shows that the sum of two trapezoidal fuzzy variables $\xi_1 = (a_1, a_2, a_3, a_4)$ and $\xi_2 = (b_1, b_2, b_3, b_4)$ is also a trapezoidal fuzzy variable, and $\xi_1 + \xi_2 = (a_1 + b_1, a_2 + b_2, a_3 + b_3, a_4 + b_4)$.

Similarly, according to Theorem 3.4, we can calculate that the membership function μ of the product of a trapezoidal fuzzy variable $\xi = (a_1, a_2, a_3, a_4)$ and a scalar number ω is

$$\mu_{\omega\xi}(t) = \sup\{\mu_\xi(t_1) | t = \omega t_1\}$$

which produces that

$$\omega \cdot \xi = \begin{cases} (\omega a_1, \omega a_2, \omega a_3, \omega a_4), & \text{if } \omega \ge 0 \\ (\omega a_4, \omega a_3, \omega a_2, \omega a_1), & \text{if } \omega < 0. \end{cases}$$

That is, the product of a scalar number ω and a trapezoidal fuzzy variable $\xi = (a_1, a_2, a_3, a_4)$ is also a trapezoidal fuzzy variable.

Note that the *triangular fuzzy variable* $\xi = (r_1, r_2, r_4)$ is in fact a special *trapezoidal fuzzy variable* $\xi = (r_1, r_2, r_3, r_4)$ when $r_2 = r_3$. Therefore, we can easily get that the sum of two triangular fuzzy variables $\xi_1 = (a_1, a_2, a_3)$ and $\xi_2 = (b_1, b_2, b_3)$ is also a triangular fuzzy variable, and $\xi_1 + \xi_2 = (a_1 + b_1, a_2 + b_2, a_3 + b_3$. The product of a scalar number ω and a triangular fuzzy variable $\xi = (a_1, a_2, a_3)$ is also a triangular fuzzy variable, and

$$\omega \cdot \xi = \begin{cases} (\omega a_1, \omega a_2, \omega a_3), & \text{if } \omega \geq 0 \\ (\omega a_3, \omega a_2, \omega a_1), & \text{if } \omega < 0. \end{cases}$$

Example 3.11. Let $\xi_1 \sim \mathcal{N}(e_1, \sigma_1)$ and $\xi_2 \sim \mathcal{N}(e_2, \sigma_2)$ be two normal fuzzy variables. It can be proven that for any real numbers ω_1 and ω_2, the fuzzy variable $\omega_1 \xi_1 + \omega_2 \xi_2$ is also a normal fuzzy variable whose membership function is

$$\mu(t) = 2\left(1 + \exp\left(\frac{\pi|t - (\omega_1 e_1 + \omega_2 e_2)|}{\sqrt{6}(|\omega_1|\sigma_1 + |\omega_2|\sigma_2)}\right)\right)^{-1}, \quad t \in R.$$

Example 3.12. Let $\xi_1 = (a_1, a_2)$ and $\xi_2 = (b_1, b_2)$ be two equipossible fuzzy variables. It can be proven that the fuzzy variable $\xi_1 + \xi_2$ is also an equipossible fuzzy variable, and

$$\xi_1 + \xi_2 = (a_1 + b_1, a_2 + b_2).$$

Their product $\xi_1 \cdot \xi_2$ is also an equipossible fuzzy variable, and

$$\xi_1 \cdot \xi_2 = \left(\min_{a_1 \leq y \leq a_2, b_1 \leq z \leq b_2} yz, \max_{a_1 \leq y \leq a_2, b_1 \leq z \leq b_2} yz\right).$$

Example 3.13. Let $\xi_1, \xi_2, \cdots, \xi_n$ be independent fuzzy variables with membership functions $\mu_1, \mu_2, \cdots, \mu_n$, respectively, and $\Re^n \to \Re$ be a function. Then for any set B of real numbers, the credibility $\text{Cr}\{f(t_1, t_2, \cdots, t_n) \in B\}$ is

$$\frac{1}{2}\left(\sup_{f(t_1,t_2,\cdots,t_n) \in B} \min_{1 \leq i \leq n} \mu_i(t_i) + 1 - \sup_{f(t_1,t_2,\cdots,t_n) \in B^c} \min_{1 \leq i \leq n} \mu_i(t_i)\right).$$

Expected Value

Expected value operator calculates the average value of a fuzzy variable.

Definition 3.9. *(Liu and Liu [55]) Let ξ be a fuzzy variable. Then the expected value of ξ is defined by*

$$E[\xi] = \int_0^{+\infty} \text{Cr}\{\xi \geq t\}\text{d}t - \int_{-\infty}^0 \text{Cr}\{\xi \leq t\}\text{d}t \tag{3.17}$$

provided that at least one of the two integrals is finite.

Example 3.14. Let $\xi = (r_1, r_2, r_3)$ be the triangular fuzzy variable. We know from the credibility inversion theorem that

$$\mathrm{Cr}\{\xi \le t\} = \begin{cases} 1, & r_3 \le t \\ \dfrac{r_3 - 2r_2 + t}{2(r_3 - r_2)}, & r_2 \le t \le r_3 \\ \dfrac{t - r_1}{2(r_2 - r_1)}, & r_1 \le t \le r_2 \\ 0, & \text{otherwise,} \end{cases}$$

and

$$\mathrm{Cr}\{\xi \ge t\} = \begin{cases} 0, & r_3 \le t \\ \dfrac{r_3 - t}{2(r_3 - r_2)}, & r_2 \le t \le r_3 \\ \dfrac{2r_2 - r_1 - t}{2(r_2 - r_1)}, & r_1 \le t \le r_2 \\ 1, & \text{otherwise.} \end{cases}$$

Thus, if $0 \le r_1 < r_2 < r_3$, we have $\mathrm{Cr}\{\xi \le t\} \equiv 0$ when $t < 0$. Then

$$\begin{aligned} E[\xi] = & \left(\int_0^{r_1} 1\mathrm{d}t + \int_{r_1}^{r_2} \frac{2r_2 - r_1 - t}{2(r_2 - r_1)}\mathrm{d}t + \int_{r_2}^{r_3} \frac{r_3 - t}{2(r_3 - r_2)}\mathrm{d}t \right.\\ & \left. + \int_{r_3}^{+\infty} 0\mathrm{d}t\right) - \int_{-\infty}^{0} 0\mathrm{d}t = \frac{1}{4}(r_1 + 2r_2 + r_3). \end{aligned}$$

If $r_1 < 0 \le r_2$, then

$$\begin{aligned} E[\xi] = & \left(\int_0^{r_2} \frac{2r_2 - r_1 - t}{2(r_2 - r_1)}\mathrm{d}t + \int_{r_2}^{r_3} \frac{r_3 - t}{2(r_3 - r_2)}\mathrm{d}t + \int_{r_3}^{+\infty} 0\mathrm{d}t\right) \\ & - \left(\int_{-\infty}^{r_1} 0\mathrm{d}t + \int_{r_1}^{0} \frac{t - r_1}{2(r_2 - r_1)}\mathrm{d}t\right) = \frac{1}{4}(r_1 + 2r_2 + r_3). \end{aligned}$$

If $r_1 < r_2 < 0 < r_3$, then

$$\begin{aligned} E[\xi] = & \left(\int_0^{r_3} \frac{r_3 - t}{2(r_3 - r_2)}\mathrm{d}t + \int_{r_3}^{+\infty} 0\mathrm{d}t\right) - \left(\int_{-\infty}^{r_1} 0\mathrm{d}t + \right. \\ & \left. \int_{r_1}^{r_2} \frac{t - r_1}{2(r_2 - r_1)}\mathrm{d}t + \int_{r_2}^{0} \frac{r_3 - 2r_2 + t}{2(r_3 - r_2)}\mathrm{d}t\right) = \frac{1}{4}(r_1 + 2r_2 + r_3). \end{aligned}$$

If $r_1 < r_2 < r_3 \le 0$, then

$$E[\xi] = \int_0^{+\infty} 0\mathrm{d}t - \left(\int_{-\infty}^{r_1} 0\mathrm{d}t + \int_{r_1}^{r_2} \frac{t-r_1}{2(r_2-r_1)}\mathrm{d}t + \int_{r_2}^{r_3} \frac{r_3-2r_2+t}{2(r_3-r_2)}\mathrm{d}t + \int_{r_3}^{0} 1\mathrm{d}t\right) = \frac{1}{4}(r_1+2r_2+r_3).$$

Therefore, the expected value of the triangular fuzzy variable $\xi = (r_1, r_2, r_3)$ is always

$$E[\xi] = \frac{1}{4}(r_1+2r_2+r_3). \tag{3.18}$$

Example 3.15. The expected value of a trapezoidal fuzzy variable $\xi = (r_1, r_2, r_3, r_4)$ is

$$E[\xi] = \frac{1}{4}(r_1+r_2+r_3+r_4).$$

Example 3.16. The expected value of a normal fuzzy variable $\xi \sim \mathcal{N}(e, \sigma)$ is

$$E[\xi] = e.$$

Example 3.17. The expected value of an equipossible fuzzy variable $\xi = (r_1, r_2)$ is

$$E[\xi] = (r_2+r_1)/2.$$

Theorem 3.5 *(Liu and Liu [57]) Let ξ_1 and ξ_2 be independent fuzzy variables with finite expected values. Then for any numbers a_1 and a_2, we have*

$$E[a_1\xi_1 + a_2\xi_2] = a_1E[\xi_1] + a_2E[\xi_2]. \tag{3.19}$$

Variance

Definition 3.10 *(Liu and Liu [55]) Let ξ be a fuzzy variable with finite expected value e. Then the variance of ξ is defined by*

$$V[\xi] = E[(\xi - e)^2].$$

Example 3.18. Let ξ be an equipossible fuzzy variable (a, b). Remember that $E[\xi] = e = (a+b)/2$. Then for any positive number t, we have

$$\mathrm{Cr}\{(\xi-e)^2 \ge t\} = \begin{cases} 1/2, & \text{if } t \le (b-a)^2/4 \\ 0, & \text{if } t > (b-a)^2/4. \end{cases}$$

Thus the variance is

$$V[\xi] = \int_0^{+\infty} \mathrm{Cr}\{(\xi-e)^2 \ge t\}\mathrm{d}t = \int_0^{(b-a)^2/4} \frac{1}{2}\mathrm{d}t = \frac{(b-a)^2}{8}.$$

Example 3.19. Let $\xi = (r_1, r_2, r_3)$ be a triangular fuzzy variable. Then its variance is

$$V[\xi] = \frac{33\alpha^3 + 21\alpha^2\beta + 11\alpha\beta^2 - \beta^3}{384\alpha},$$

where $\alpha = \max\{r_2 - r_1, r_3 - r_2\}$ and $\beta = \min\{r_2 - r_1, r_3 - r_2\}$. Especially, when $\xi = (r_1, r_2, r_3)$ is a symmetric triangular fuzzy variable, i.e., $r_3 - r_2 = r_2 - r_1$, its variance is

$$V[\xi] = (r_3 - r_1)^2/24.$$

Example 3.20. Let $\xi = (r_1, r_2, r_3, r_4)$ be a symmetric trapezoidal fuzzy variable, i.e., $r_4 - r_3 = r_2 - r_1$. Then its variance is

$$V[\xi] = ((r_4 - r_1)^2 + (r_4 - r_1)(r_3 - r_2) + (r_3 - r_2)^2)/24.$$

Example 3.21. Let $\xi \sim \mathcal{N}(e, \sigma)$ be a normal fuzzy variable. Then its variance is

$$V[\xi] = \sigma^2.$$

Theorem 3.6 *(Liu [58]) Let a and b be real numbers and ξ a fuzzy variable whose variance exists. Then*

$$V[a\xi + b] = a^2 V[\xi]. \tag{3.20}$$

Example 3.22. Let $\xi_1 \sim \mathcal{N}(e_1, \sigma_1)$ and $\xi_2 \sim \mathcal{N}(e_2, \sigma_2)$ be two normal fuzzy variables, and a_1 and a_2 any real numbers. Then

$$E[a_1\xi_1 + a_2\xi_2] = a_1 e_1 + a_2 e_2 \quad \text{and}$$

$$V[a_1\xi_1 + a_2\xi_2] = (|a_1|\sigma_1 + |a_2|\sigma_2)^2.$$

Semivariance

Definition 3.11 *(Huang [37]) Let ξ be a fuzzy variable with finite expected value e. Then the semivariance of ξ is defined by*

$$SV[\xi] = E[[(\xi - e)^-]^2],$$

where

$$(\xi - e)^- = \begin{cases} \xi - e, & \text{if } \xi \le e \\ 0, & \text{if } \xi > e. \end{cases} \tag{3.21}$$

Example 3.23. Let $\xi = (a, b, c)$ be a triangular fuzzy variable with $b - a > c - b$. Then the semivariance of ξ is

$$SV[\xi] = \frac{(e + a)(e^2 - a^2)}{2(b - a)} - \frac{e^3 - a^3}{3(b - a)} - \frac{ae(e - a)}{b - a}$$

where $e = (a + 2b + c)/4$.

Example 3.24. Let $\xi = (a, b, c)$ be a triangular fuzzy variable with $b - a < c - b$. Then the semivariance of ξ, i.e., $SV[\xi]$ is

$$\frac{3eb + ab - 3ae + a^2 - 2b^2}{6} + \frac{e^3 - 4b^3 + 3ce^2 - 6be^2 + 9b^2e + 3cb^2 - 6cbe}{6(c - b)}$$

where $e = (a + 2b + c)/4$.

Theorem 3.7 *(Huang [37]) Let ξ be a fuzzy variable, $SV[\xi]$ and $V[\xi]$ the semivariance and variance of ξ, respectively. Then $0 \le SV[\xi] \le V[\xi]$.*

Proof: Let e be the expected value of a fuzzy variable ξ. The nonnegativity of variance and semivariance is clear. For any real number t, we have

$$\{\theta \mid (\xi(\theta) - e)^2 \ge t\} \supset \{\theta \mid [(\xi(\theta) - e)^-]^2 \ge t\},$$

which implies that

$$\mathrm{Cr}\{(\xi - e)^2 \ge t\} \ge \mathrm{Cr}\{[(\xi - e)^-]^2 \ge t\}, \quad \forall t$$

because credibility is monotonous.

It follows from the definition of variance and semivariance that

$$V[\xi] = \int_0^{+\infty} \mathrm{Cr}\{(\xi - e)^2 \ge t\}dt \ge \int_0^{+\infty} \mathrm{Cr}\{[(\xi - e)^-]^2 \ge t\}dt = SV[\xi].$$

Theorem 3.8 *(Huang [37]) Let ξ be a fuzzy variable with symmetric membership function. Then $SV[\xi] = V[\xi]$.*

Proof: Let ξ be a fuzzy variable with symmetric membership function about its expected value e. From the definition of variance, we have

$$V[\xi] = E[(\xi - e)^2] = \int_0^{+\infty} \mathrm{Cr}\{(\xi - e)^2 \ge t\}dt.$$

Since the membership function of ξ is symmetric about e, we have

$$\mathrm{Cr}\{(\xi - e)^2 \ge t\} = \mathrm{Cr}\{[(\xi - e)^-]^2 \ge t\}, \quad \forall t.$$

Therefore,

$$V[\xi] = \int_0^{+\infty} \mathrm{Cr}\{(\xi - e)^2 \ge t\}dt = \int_0^{+\infty} \mathrm{Cr}\{[(\xi - e)^-]^2 \ge t\}dt = SV[\xi].$$

Remark 3.4. Theorem 3.8 tells us that when a membership function of portfolio return is symmetrical, the variance value and the semivariance value of the fuzzy portfolio return will be the same. However, when a membership function of a fuzzy portfolio return is asymmetrical, Theorem 3.7 tells us that the variance value and the semivariance value of the fuzzy portfolio return will be different.

β-Value

Definition 3.12 *(Liu [56]) Let ξ be a fuzzy variable, and $\beta \in (0,1]$. Then*

$$\xi_{\sup}(\beta) = \sup\left\{r \mid \mathrm{Cr}\left\{\xi \geq r\right\} \geq \beta\right\} \tag{3.22}$$

is called the β-value of ξ.

Example 3.25. Let $\xi = [a, b]$ be an equipossible fuzzy variable. Then its β-value is

$$\xi_{\sup}(\beta) = \begin{cases} b, & \text{if } \beta \leq 0.5 \\ a, & \text{if } \beta > 0.5. \end{cases}$$

Example 3.26. Let $\xi = (r_1, r_2, r_3)$ be a triangular fuzzy variable. Then its β-value is

$$\xi_{\sup}(\beta) = \begin{cases} 2\beta r_2 + (1-2\beta)r_3, & \text{if } \beta \leq 0.5 \\ (2\beta-1)r_1 + (2-2\beta)r_2, & \text{if } \beta > 0.5. \end{cases}$$

Example 3.27. Let $\xi = (r_1, r_2, r_3, r_4)$ be a trapeziodal fuzzy variable. Then its β-value is

$$\xi_{\sup}(\beta) = \begin{cases} 2\beta r_3 + (1-2\beta)r_4, & \text{if } \beta \leq 0.5 \\ (2\beta-1)r_1 + (2-2\beta)r_2, & \text{if } \beta > 0.5. \end{cases}$$

Example 3.28. Let $\xi \sim \mathcal{N}(e, \sigma)$ be a normal fuzzy variable. Then its β-value is

$$\xi_{\sup}(\beta) = e - \frac{\sqrt{6}\sigma}{\pi} \ln \frac{\beta}{1-\beta}.$$

Theorem 3.9 *(Liu [58]). Let $\xi_{\sup}(\beta)$ be the β-value of the fuzzy variable ξ. Then $\xi_{\sup}(\beta)$ is a decreasing and left-continuous function of β.*

Theorem 3.10. *Let $\xi_{\sup}(\beta)$ be the β-value of the fuzzy variable ξ. Then if $\lambda \geq 0$, we have $(\lambda\xi)_{\sup}(\beta) = \lambda\xi_{\sup}(\beta)$.*

Theorem 3.11 *(Li and Liu [49]). Let ξ and η be two independent fuzzy variables. Then for any $\beta \in (0,1]$, we have*

$$(\xi + \eta)_{\sup}(\beta) = \xi_{\sup}(\beta) + \eta_{\sup}(\beta) \tag{3.23}$$

Proof: According to monotonicity property of credibility measure, for any $\epsilon > 0$, we have

$$\begin{aligned} &\mathrm{Cr}\{\xi + \eta \geq \xi_{\sup}(\beta) + \eta_{\sup}(\beta) - \epsilon\} \\ &\geq \mathrm{Cr}\Big\{\{\xi \geq \xi_{\sup}(\beta) - \epsilon/2\} \cap \{\eta \geq \eta_{\sup}(\beta) - \epsilon/2\}\Big\}. \end{aligned}$$

Since ξ and η are independent fuzzy variables, according to Definition 3.7, we have

$$\begin{aligned}&\mathrm{Cr}\{\xi+\eta\geq\xi_{\sup}(\beta)+\eta_{\sup}(\beta)-\epsilon\}\\&\geq\mathrm{Cr}\Big\{\{\xi\geq\xi_{\sup}(\beta)-\epsilon/2\}\cap\{\eta\geq\eta_{\sup}(\beta)-\epsilon/2\}\Big\}\\&=\mathrm{Cr}\{\xi\geq\xi_{\sup}(\beta)-\epsilon/2\}\wedge\mathrm{Cr}\{\eta\geq\eta_{\sup}(\beta)-\epsilon/2\}\geq\beta\end{aligned}$$

which implies that

$$(\xi+\eta)_{\sup}(\beta)\geq\xi_{\sup}(\beta)+\eta_{\sup}(\beta)-\epsilon. \tag{3.24}$$

According to monotonicity property of credibility measure, for any $\epsilon>0$, we have

$$\begin{aligned}&\mathrm{Cr}\{\xi+\eta\geq\xi_{\sup}(\beta)+\eta_{\sup}(\beta)+\epsilon\}\\&\leq\mathrm{Cr}\Big\{\{\xi\geq\xi_{\sup}(\beta)+\epsilon/2\}\cup\{\eta\geq\eta_{\sup}(\beta)+\epsilon/2\}\Big\}.\end{aligned}$$

Since ξ and η are independent fuzzy variables, according to Theorem 3.3, we have

$$\begin{aligned}&\mathrm{Cr}\{\xi+\eta\geq\xi_{\sup}(\beta)+\eta_{\sup}(\beta)+\epsilon\}\\&\leq\mathrm{Cr}\Big\{\{\xi\geq\xi_{\sup}(\beta)+\epsilon/2\}\cup\{\eta\geq\eta_{\sup}(\beta)+\epsilon/2\}\Big\}\\&=\mathrm{Cr}\{\xi\geq\xi_{\sup}(\beta)+\epsilon/2\}\vee\mathrm{Cr}\{\eta\geq\eta_{\sup}(\beta)+\epsilon/2\}<\beta\end{aligned}$$

which implies that

$$(\xi+\eta)_{\sup}(\beta)\leq\xi_{\sup}(\beta)+\eta_{\sup}(\beta)+\epsilon. \tag{3.25}$$

It follows from (3.24) and (3.25) that

$$\xi_{\sup}(\beta)+\eta_{\sup}(\beta)+\epsilon\geq(\xi+\eta)_{\sup}(\beta)\geq\xi_{\sup}(\beta)+\eta_{\sup}(\beta)-\epsilon.$$

Letting $\epsilon\to 0$, we have

$$(\xi+\eta)_{\sup}(\beta)=\xi_{\sup}(\beta)+\eta_{\sup}(\beta).$$

Entropy

Fuzzy entropy is a measure of fuzzy uncertainty. It measures the difficulty degree of predicting the specific value that a fuzzy variable will take.

Definition 3.13 *(Li and Liu [51]). Let ξ be a fuzzy variable with continuous membership function. Then its entropy is defined by*

$$H[\xi]=\int_{-\infty}^{\infty}S(\mathrm{Cr}\{\xi=t\})\mathrm{d}t \tag{3.26}$$

where $S(y)=-y\ln y-(1-y)\ln(1-y)$.

Since for any fuzzy variable ξ with continuous membership function μ, we have $\mathrm{Cr}\{\xi = r\} = \dfrac{\mu(r)}{2}$ for each $r \in R$. Thus, the entropy can be expressed by

$$H[\xi] = -\int_{-\infty}^{\infty} \left(\frac{\mu(r)}{2} \ln \frac{\mu(r)}{2} + \left(1 - \frac{\mu(r)}{2}\right) \ln \left(1 - \frac{\mu(r)}{2}\right)\right) \mathrm{d}r. \quad (3.27)$$

Example 3.29. Let ξ be a triangular fuzzy variable $\xi = (r_1, r_2, r_3)$. Then its entropy is $H[\xi] = (r_3 - r_1)/2$.

Example 3.30. Let ξ be a trapezoidal fuzzy variable $\xi = (r_1, r_2, r_3, r_4)$. Then its entropy is $H[\xi] = (r_4 - r_1)/2 + (\ln 2 - 0.5)(r_3 - r_2)$.

Example 3.31. Let $\xi \sim \mathcal{N}(e, \sigma)$ be a normal fuzzy variable. Then its entropy is $H[\xi] = \sqrt{6}\pi\sigma/3$.

Remark 3.5. Let ξ be a fuzzy variable with continuous membership function and taking continuous values in the interval $[a, b]$. Then we can find that $H[\xi] \leq (b-a) \ln 2$ and that the equality holds if and only if ξ is an equipossible fuzzy variable in the interval $[a, b]$. Since a fuzzy variable with maximum entropy distributes most dispersively and it will be most difficult to predict whether this fuzzy variable will take the specific value, for the safety reason of decision making, if the investors can only give the interval that a security return may lie in and nothing else, they can use the equipossible fuzzy variable to describe this security return.

Remark 3.6. Let ξ be a fuzzy variable with a continuous membership function and having finite expected value e and variance value σ^2. It has been proven [50] that $H[\xi] \leq \sqrt{6}\pi\sigma/3$ and that the equality holds if ξ is a normal fuzzy variable with expected value e and variance σ^2. Since a fuzzy variable with maximum entropy distributes most dispersively and it will be most difficult to predict whether this fuzzy variable will take the specific value, for the safety reason of decision making, if the investors can predict only the expected value and variance value of a security return and nothing else, for the safety reason of decision making, they can use the normal fuzzy variable to describe the security return.

3.2 Mean-Risk Model

In reality, some people do not like taking plane because when the plane crashes, it is almost sure that people in the plane will lose their lives though the chance of crashing event is very low. This phenomenon implies that when judging if an event is risky or not, people will consider both the occurrence chance and the severity level of the bad event. This is also true in portfolio investment. To give an instinct information about each likely loss and the corresponding occurrence chance of the loss for portfolio investment with fuzzy returns, Huang [38] defined the concept of risk curve.

3.2.1 Risk Curve

Definition 3.14 *(Huang [38]) Let ξ denote the fuzzy return of a portfolio, and r_f the risk-free interest rate. Then the curve*

$$R(r) = \mathrm{Cr}\{r_f - \xi \geq r\}, \quad \forall r \geq 0 \tag{3.28}$$

is called the risk curve of the portfolio, and r the loss severity indicator.

It is easy to see that $r_f - \xi$ is the deviation of the portfolio return from the risk-free interest rate when $r_f - \xi \geq 0$. Then the value $r_f - \xi$ can easily be understood as a loss. Since the portfolio return is variable, the loss value $r_f - \xi$ may be any non-negative values which can be expressed by

$$r_f - \xi \geq r, \quad r \geq 0.$$

Please note that r is not one specific number but any non-negative numbers, so $r_f - \xi \geq r$ describes all the likely losses of the portfolio, and the curve $R(r)$ gives corresponding occurrence credibility levels of all these losses.

Example 3.32. Let $\xi(a_1, a_1, a_3)$ denote a triangular fuzzy portfolio return. Then risk curve of ξ is as follows,

$$R(r) = \mathrm{Cr}\{r_f - \xi \geq r\} = \begin{cases} 1, & \text{if } r_f - a_3 > r \geq 0 \\ \dfrac{a_3 - 2a_2 + r_f - r}{2(a_3 - a_2)}, & \text{if } r_f - a_2 > r \geq r_f - a_3 \\ \dfrac{r_f - a_1 - r}{2(a_2 - a_1)}, & \text{if } r_f - a_1 > r \geq r_f - a_2 \\ 0, & \text{otherwise.} \end{cases} \tag{3.29}$$

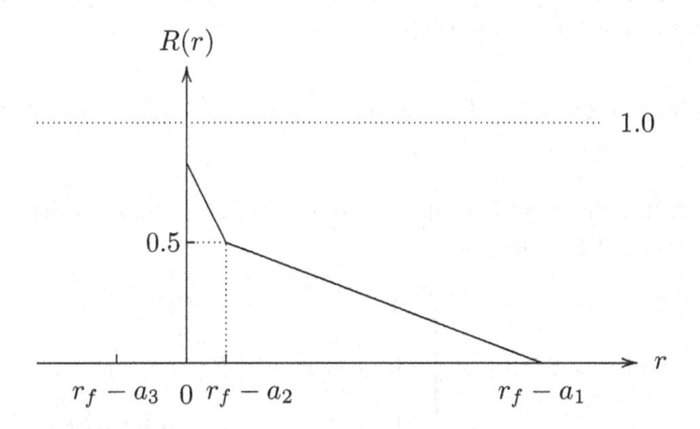

Fig. 3.8 Risk curve of a portfolio with triangular fuzzy return.

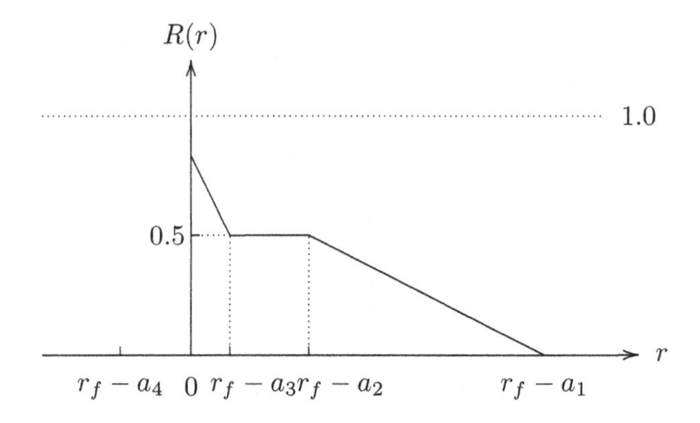

Fig. 3.9 Risk curve of a portfolio with trapezoidal fuzzy return.

Example 3.33. Let $\xi(a_1, a_2, a_3, a_4)$ denote a trapezoidal fuzzy portfolio return. Then risk curve of ξ is as follows,

$$R(r) = \mathrm{Cr}\{r_f - \xi \geq r\} = \begin{cases} 1, & \text{if } r_f - a_4 > r \geq 0 \\ \dfrac{a_4 - 2a_3 + r_f - r}{2(a_4 - a_3)}, & \text{if } r_f - a_3 > r \geq r_f - a_4 \\ 0.5, & \text{if } r_f - a_2 > r \geq r_f - a_3 \\ \dfrac{r_f - a_1 - r}{2(a_2 - a_1)}, & \text{if } r_f - a_1 > r \geq r_f - a_2 \\ 0, & \text{otherwise.} \end{cases} \tag{3.30}$$

Example 3.34. Let $\xi \sim \mathcal{N}(e, \sigma)$ denote a normal fuzzy portfolio return. Then risk curve of ξ is as follows,

$$R(r) = \mathrm{Cr}\{r_f - \xi \geq r\} = \left(1 + \exp\left(\frac{\pi(e - r_f + r)}{\sqrt{6}\sigma}\right)\right)^{-1}, \quad r \geq 0. \tag{3.31}$$

Example 3.35. Let $\xi = (a, b)$ denote an equipossible fuzzy portfolio return. Then risk curve of ξ is as follows,

$$R(r) = \mathrm{Cr}\{r_f - \xi \geq r\} = \begin{cases} 1, & \text{if } r \leq r_f - b \\ 0.5, & \text{if } r_f - b \leq r \leq r_f - a \\ 0, & \text{otherwise.} \end{cases} \tag{3.32}$$

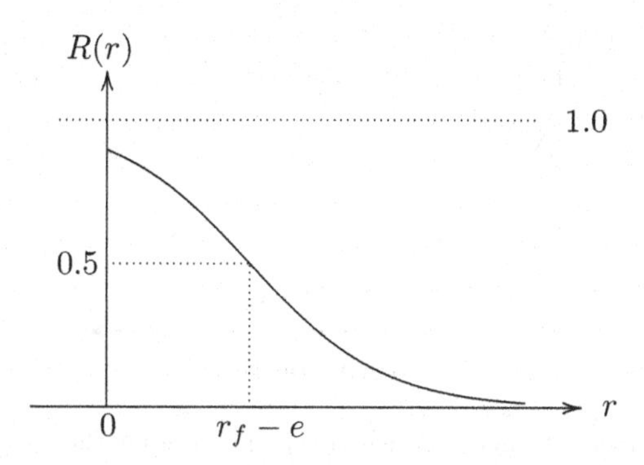

Fig. 3.10 Risk curve of a portfolio with normal fuzzy return.

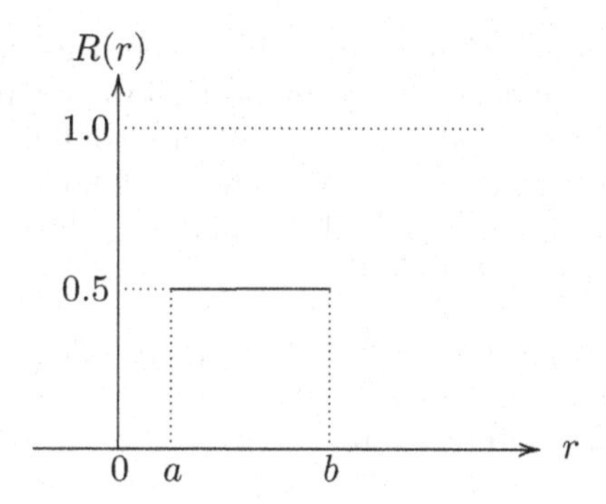

Fig. 3.11 Risk curve of a portfolio with equipossible fuzzy return.

3.2.2 Confidence Curve and Safe Portfolio

To determine a level of a risk, according to Definition 3.14, three inputs should be given. First input is the value r, the loss severity level. Second input is the occurrence chance of the loss event, i.e., $\mathrm{Cr}\{r_f - \xi \geq r\}$. Third input is the investors' subjective judgement to the above two inputs. Different investors have different judgements. Given any value r, an investor should be able to give his/her maximal tolerance towards the occurrence chance of the loss being equal to or greater than r by answering what-if questions in Table 2.5. In fuzzy portfolio selection, occurrence chance of a fuzzy event is measured by credibility value. We call the curve the confidence curve $\alpha(r)$ that gives the investor's maximal tolerance towards the occurrence chances of all the potential losses. Though different investors have different confidence curves, the common trend of the curve is that the severer the loss, the lower the tolerance of occurrence chance of the loss. The general trend of the confidence curve is given in Fig.3.12. Three examples of confidence curve are presented in Subsection 2.2.2 in Chapter 2.

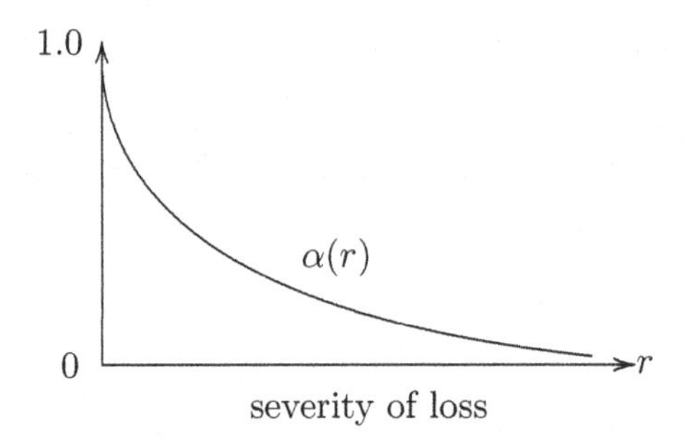

Fig. 3.12 General trend of a confidence curve: The higher the loss value, the lower the tolerance of the occurrence chance of the loss.

It is easy to understand that a portfolio is safe if its risk curve is totally below the investor's confidence curve. A portfolio is regarded to be risky if any part of its risk curve is above the investor's confidence curve (see Fig. 3.12). The mathematical expression of a safe portfolio is as follows:

Let ξ be the fuzzy return of a portfolio A, and $\alpha(r)$ the investor's confidence curve. We say A is a safe portfolio if

$$R(r) = \mathrm{Cr}\{(r_f - \xi) \geq r\} \leq \alpha(r), \quad \forall r \geq 0,$$

where r_f is the risk-free interest rate.

3.2.3 Mean-Risk Model

Let x_i denote the investment proportions in securities i, and ξ_i the i-th security returns which are fuzzy. According to Definition 3.14, the risk curve of a portfolio $(x_1, x_2, \cdots, x_n)$ is

$$R(x_1, x_2, \cdots, x_n; r) = \mathrm{Cr}\left\{r_f - (\xi_1 x_1 + \xi_2 x_2 + \cdots + \xi_n x_n) \geq r\right\}.$$

Let $\alpha(r)$ be an investor's confidence curve. The philosophy of mean-risk model is to pursue maximum expected return among the safe portfolios whose risk curves are below the investor's confidence curve. To express it in mathematical way, we have the model as follows:

$$\begin{cases} \max E[\xi_1 x_1 + \xi_2 x_2 + \cdots + \xi_n x_n] \\ \text{subject to:} \\ \qquad R(x_1, x_2, \cdots, x_n; r) \leq \alpha(r), \ \forall r \geq 0 \\ \qquad x_1 + x_2 + \cdots + x_n = 1 \\ \qquad x_i \geq 0, \quad i = 1, 2, \cdots, n. \end{cases} \tag{3.33}$$

The constraint $R(x_1, x_2, \cdots, x_n; r) \le \alpha(r)$ requires that the credibility value of each likely loss of a selected portfolio must be lower than the investor's tolerance level. The constraint $x_i \ge 0$ implies that short sales are not allowed in the investment.

3.2.4 Crisp Equivalent

One way of solving the mean-risk model is to convert the expected value and risk curve of the portfolio into their crisp equivalents and use traditional methods to solve the mean-risk model. Luckily, for independent fuzzy security returns, we have the transformation theorem as follows:

Theorem 3.12 *Let $\xi_1, \xi_2, \cdots, \xi_n$ be independent fuzzy variables with continuous credibility distributions $\Phi_1, \Phi_2, \cdots, \Phi_n$, respectively. If*

$$\lim_{t \to -\infty} \Phi_i(t) = 0, \quad \lim_{t \to \infty} \Phi_i(t) = 1, \quad \textit{for} \quad i = 1, 2, \cdots, n,$$

and $\Phi_1^{-1}(\alpha), \Phi_2^{-1}(\alpha), \cdots, \Phi_n^{-1}(\alpha)$ are unique for each $\alpha \in (0,1)$, then for any $\alpha \in (0,1)$, we have

$$\Psi^{-1}(\alpha) = \Phi_1^{-1}(\alpha) + \Phi_2^{-1}(\alpha) + \cdots + \Phi_n^{-1}(\alpha), \quad 0 < \alpha < 1 \tag{3.34}$$

where Ψ is the distribution function of fuzzy variable $\xi = \xi_1 + \xi_2 + \cdots + \xi_n$.

Proof: According to monotonicity property of credibility measure, for any given $\alpha \in (0,1)$, we have

$$\mathrm{Cr}\left\{\sum_{i=1}^{n} \xi_i \le \sum_{i=1}^{n} \Phi_i^{-1}(\alpha)\right\} \ge \mathrm{Cr}\left\{\bigcap_{i=1}^{n} \left(\xi_i \le \Phi_i^{-1}(\alpha)\right)\right\}.$$

Since $\xi_1, \xi_2, \cdots, \xi_n$ are independent fuzzy variables, according to Equation (3.13), we have

$$\begin{aligned}
\mathrm{Cr}\left\{\sum_{i=1}^{n} \xi_i \le \sum_{i=1}^{n} \Phi_i^{-1}(\alpha)\right\} &\ge \mathrm{Cr}\left\{\bigcap_{i=1}^{n} \left(\xi_i \le \Phi_i^{-1}(\alpha)\right)\right\} \\
&= \min_{1 \le i \le n} \mathrm{Cr}\{\xi_i \le \Phi_i^{-1}(\alpha)\} = \min_{1 \le i \le n} \alpha = \alpha.
\end{aligned}$$

On the other hand, for any number $\epsilon > 0$, we have

$$\mathrm{Cr}\left\{\sum_{i=1}^{n} \xi_i \le \sum_{i=1}^{n} \Phi_i^{-1}(\alpha) - \epsilon\right\} \le \mathrm{Cr}\left\{\bigcup_{i=1}^{n} \left(\xi_i \le \Phi_i^{-1}(\alpha) - \frac{\epsilon}{n}\right)\right\}$$

because credibility measure is monotonous. Since $\xi_1, \xi_2, \cdots, \xi_n$ are independent fuzzy variables, according to Equation (3.14), we have

$$\mathrm{Cr}\left\{\sum_{i=1}^{n}\xi_i \le \sum_{i=1}^{n}\Phi_i^{-1}(\alpha)-\epsilon\right\} \le \mathrm{Cr}\left\{\bigcup_{i=1}^{n}\left(\xi_i \le \Phi_i^{-1}(\alpha)-\frac{\epsilon}{n}\right)\right\}$$
$$= \max_{1\le i\le n}\mathrm{Cr}\left\{\xi_i \le \Phi_i^{-1}(\alpha)-\frac{\epsilon}{n}\right\} < \max_{1\le i\le n}\alpha = \alpha.$$

It follows from the continuity of credibility distributions that

$$\mathrm{Cr}\{\xi_1+\xi_2+\cdots+\xi_n \le \Phi_1^{-1}(\alpha)+\Phi_2^{-1}(\alpha)+\cdots+\Phi_n^{-1}(\alpha)\} = \alpha$$

which implies that

$$\Psi^{-1}(\alpha) = \Phi_1^{-1}(\alpha)+\Phi_2^{-1}(\alpha)+\cdots+\Phi_n^{-1}(\alpha).$$

The theorem is proven.

Theorem 3.13. *Let Φ_i denote the credibility distributions of the i-th fuzzy security return rates $\xi_i, i = 1, 2, \cdots, n$, respectively. Then the mean-risk model (3.33) can be transformed into the following linear model:*

$$\begin{cases} \max x_1E[\xi_1]+x_2E[\xi_2]+\cdots+x_nE[\xi_n] \\ \textit{subject to:} \\ \qquad x_1\Phi_1^{-1}\Big(\alpha(r)\Big)+x_2\Phi_2^{-1}\Big(\alpha(r)\Big)+\cdots+x_n\Phi_n^{-1}\Big(\alpha(r)\Big) \ge r_f - r,\ \forall r\ge 0 \\ \qquad x_1+x_2+\cdots+x_n = 1 \\ \qquad x_i \ge 0, \quad i = 1,2,\cdots,n. \end{cases} \tag{3.35}$$

Proof: It follows from linearity property of expected value that the objective function of Model (3.33) can be transformed into the objective function of Model (3.35).

It follows from Theorem 3.12 that the risk curve in Model (3.33) can be transformed into the following linear form

$$R^{-1}(x_1,x_2,\cdots,x_n;r) = x_1\Phi_1^{-1}\Big(\alpha(r)\Big)+x_2\Phi_2^{-1}\Big(\alpha(r)\Big)+\cdots+x_n\Phi_n^{-1}\Big(\alpha(r)\Big).$$

It follows from monotonicity of credibility measure that

$$x_1\Phi_1^{-1}\Big(\alpha(r)\Big)+x_2\Phi_2^{-1}\Big(\alpha(r)\Big)+\cdots+x_n\Phi_n^{-1}\Big(\alpha(r)\Big) \ge r_f - r.$$

Example 3.36. Suppose the return rates of the i-th securities are all triangular fuzzy variables $\xi_i = (a_i, b_i, c_i), i = 1, 2, \cdots, n$, respectively. Then the fuzzy mean-risk model can be transformed into the following form:

$$\begin{cases} \max \sum_{i=1}^{n} (a_i x_i + 2b_i x_i + c_i x_i) \\ \text{subject to:} \\ \quad \Big(2\alpha(r) - 1\Big) \sum_{i=1}^{n} c_i x_i + \Big(2 - 2\alpha(r)\Big) \sum_{i=1}^{n} b_i x_i \geq r_f - r, \text{ if } \alpha(r) > 0.5 \\ \quad 2\alpha(r) \sum_{i=1}^{n} b_i x_i - \Big(2\alpha(r) - 1\Big) \sum_{i=1}^{n} a_i x_i \geq r_f - r, \text{ if } \alpha(r) \leq 0.5 \\ \quad x_1 + x_2 + \cdots + x_n = 1 \\ \quad x_i \geq 0, \quad i = 1, 2, \cdots, n. \end{cases} \tag{3.36}$$

Since all security return rates are triangular fuzzy variables, the portfolio return rate is still a triangular fuzzy variable, i.e.,

$$\xi = \sum_{i=1}^{n} \xi_i x_i = \left(\sum_{i=1}^{n} a_i x_i, \sum_{i=1}^{n} b_i x_i, \sum_{i=1}^{n} c_i x_i \right).$$

Thus, we can get Model (3.36) easily.

Example 3.37. Suppose the return rates of the i-th securities are all normal fuzzy variables $\xi_i \sim \mathcal{N}(e_i, \sigma_i), i = 1, 2, \cdots, n$, respectively. Then the fuzzy mean-risk model can be transformed into the following form:

$$\begin{cases} \max e_1 x_1 + e_2 x_2 + \cdots + e_n x_n \\ \text{subject to:} \\ \quad \sum_{i=1}^{n} \left(e_i - \frac{\sqrt{6}\sigma_i}{\pi} \ln \frac{1 - \alpha(r)}{\alpha(r)} \right) x_i \geq r_f - r, \quad \forall r \geq 0 \\ \quad x_1 + x_2 + \cdots + x_n = 1 \\ \quad x_i \geq 0, \quad i = 1, 2, \cdots, n. \end{cases} \tag{3.37}$$

Since all security return rates are normal fuzzy variables, the portfolio return rate is still a normal fuzzy variable, i.e.,

$$\xi = \sum_{i=1}^{n} \xi_i x_i = \left(\sum_{i=1}^{n} e_i x_i, \sum_{i=1}^{n} \sigma_i x_i \right).$$

Thus, we can get Model (3.37) easily.

Example 3.38. Suppose the return rates of the i-th securities are normal fuzzy variables $\xi_i \sim \mathcal{N}(e_i, \sigma_i), i = 1, 2, \cdots, m$, and the return rates of the j-th securities are triangular fuzzy variables $\xi_j = (a_j, b_j, c_j), j = m + 1, m +$

$2, \cdots, n$, respectively. Then the fuzzy mean-risk model can be transformed into the following form:

$$\begin{cases} \max \sum_{i=1}^{m} e_i x_i + \sum_{i=m+1}^{n} \frac{1}{4}(a_i x_i + 2b_i x_i + c_i x_i) \\ \text{subject to:} \\ \qquad \sum_{i=m+1}^{m} \left(e_i - \frac{\sqrt{6}\sigma_i}{\pi} \ln \frac{1-\alpha(r)}{\alpha(r)} \right) x_i + \left(2\alpha(r) - 1\right) \sum_{i=m+1}^{n} c_i x_i + \\ \qquad \left(2 - 2\alpha(r)\right) \sum_{i=m+1}^{n} b_i x_i \geq r_f - r, \text{ if } \alpha(r) > 0.5 \\ \qquad \sum_{i=1}^{m} \left(e_i - \frac{\sqrt{6}\sigma_i}{\pi} \ln \frac{1-\alpha(r)}{\alpha(r)} \right) x_i + 2\alpha(r) \sum_{i=m+1}^{n} b_i x_i - \\ \qquad \left(2\alpha(r) - 1\right) \sum_{i=m+1}^{n} a_i x_i \geq r_f - r, \text{ if } \alpha(r) \leq 0.5 \\ \qquad x_1 + x_2 + \cdots + x_n = 1 \\ \qquad x_i \geq 0, \quad i = 1, 2, \cdots, n. \end{cases} \tag{3.38}$$

3.2.5 An Example

Suppose an investor wants to choose an optimal portfolio from ten securities of which five security return rates are normal fuzzy variables and the rest five the triangular fuzzy variables. The prediction of the return rates of the ten securities is given in Table 3.1. Suppose the monthly risk-free interest rate is 0.01, and the investor gives his/her confidence curve as follows:

$$\alpha(r) = \begin{cases} -2.75r + 0.43, & 0 \leq r \leq 0.12, \\ -0.5r + 0.16, & 0.12 \leq r \leq 0.3, \\ 0.01, & r \geq 0.3. \end{cases}$$

According to the mean-risk selection idea, we build the model as follows:

$$\begin{cases} \max E[\xi_1 x_1 + \xi_2 x_2 + \cdots + \xi_{10} x_{10}] \\ \text{subject to:} \\ \qquad R(x_1, x_2, \cdots, x_{10}; r) \leq \alpha(r), \ \forall r \geq 0 \\ \qquad x_1 + x_2 + \cdots + x_{10} = 1 \\ \qquad x_i \geq 0, \quad i = 1, 2, \cdots, 10 \end{cases} \tag{3.39}$$

Table 3.1 Fuzzy Return Rates of 10 Securities

Security i	$\xi_i \sim \mathcal{N}(e_i, \sigma_i)$	Security i	$\xi_i = (a_i, b_i, c_i)$
1	$\mathcal{N}(0.034, 0.12)$	6	$(-0.06, 0.020, 0.15)$
2	$\mathcal{N}(0.033, 0.10)$	7	$(-0.10, 0.030, 0.20)$
3	$\mathcal{N}(0.039, 0.12)$	8	$(-0.12, 0.032, 0.2)$
4	$\mathcal{N}(0.028, 0.08)$	9	$(-0.20, 0.04, 0.28)$
5	$\mathcal{N}(0.025, 0.08)$	10	$(-0.16, 0.03, 0.30)$

where $R(x_1, x_2, \cdots, x_{10}; r)$ is the risk curve of the portfolio defined as

$$R(x_1, x_2, \cdots, x_{10}; r) = \text{Cr}\{0.01 - (\xi_1 x_1 + \xi_2 x_2 + \cdots + \xi_{10} x_{10}) \geq r\}.$$

According to Model (3.38), we can change Model (3.39) into the following linear programming model. Please note that $\alpha(r) < 0.5$ in the example.

$$\begin{cases} \max \sum_{i=1}^{5} e_i x_i + \sum_{i=6}^{10} \frac{1}{4}(a_i x_i + 2b_i x_i + c_i x_i) \\ \text{subject to:} \\ \qquad \sum_{i=1}^{5} \left(e_i - \frac{\sqrt{6}\sigma_i}{\pi} \ln \frac{1-\alpha(r)}{\alpha(r)} \right) x_i + 2\alpha(r) \sum_{i=6}^{10} b_i x_i - \\ \qquad \left(2\alpha(r) - 1\right) \sum_{i=6}^{10} a_i x_i \geq 0.01 - r \\ \qquad x_1 + x_2 + \cdots + x_{10} = 1 \\ \qquad x_i \geq 0, \quad i = 1, 2, \cdots, 10. \end{cases} \tag{3.40}$$

Though theoretically, when solving the mean-risk model, r should be any nonnegative numbers, in reality, r can be limited to a certain interval by analyzing the problem. In the example, since the confidence curve is a horizontal line when $r \geq 0.3$ and the risk curve is a decreasing function of r, risk curve will be below the confidence curve if $R(x_1, x_2, \cdots, x_{10}; r) \leq \alpha(r)$ holds for any $r \in [0, 0.3]$. Since risk curve is a continuous function of r, it is enough for us to check if the points on the risk curve are all lower than the points on the confidence curve for $(r = 0, \alpha = 0.43)$, $(r = 0.02, \alpha = 0.375)$, $(r = 0.04, \alpha = 0.32)$, $(r = 0.06, \alpha = 0.265), \cdots, (r = 0.3, \alpha = 0.01)$. That is, we just need to solve Model (3.41) given below. By using "Solver" in "Excel", we get the optimal portfolio shown in Table 3.2. The maximum expected return is 0.042. As shown in Fig. 3.13, risk curve of the optimal portfolio is totally below the investor's confidence curve. Given any loss value r, the loss occurrence credibility is not greater than the investor's tolerable credibility.

Table 3.2 Allocation of Money to Ten Securities

Security i	1	2	3	4	5
Allocation of money	0.00%	0.00 %	0.00%	0.00%	0.00%
Security i	6	7	8	9	10
Allocation of money	0.00%	78.57 %	0.00 %	0.00%	21.43%

Or given any occurrence credibility $\alpha(r)$, the loss level is not greater than the investor's tolerable loss level.

$$\begin{cases} \max \sum_{i=1}^{5} e_i x_i + \sum_{i=6}^{10} \frac{1}{4}(a_i x_i + 2b_i x_i + c_i x_i) \\ \text{subject to:} \\ \quad \sum_{i=1}^{5} \left(e_i - \frac{\sqrt{6}\sigma_i}{\pi} \ln \frac{1-0.43}{0.43} \right) x_i + 2 \times 0.43 \sum_{i=6}^{10} b_i x_i - \\ \quad \left(2 \times 0.43 - 1\right) \sum_{i=6}^{10} a_i x_i \geq 0.01 \\ \quad \sum_{i=1}^{5} \left(e_i - \frac{\sqrt{6}\sigma_i}{\pi} \ln \frac{1-0.375}{0.375} \right) x_i + 2 \times 0.375 \sum_{i=6}^{10} b_i x_i - \\ \quad \left(2 \times 0.375 - 1\right) \sum_{i=6}^{10} a_i x_i \geq 0.01 - 0.02 \\ \quad \sum_{i=1}^{5} \left(e_i - \frac{\sqrt{6}\sigma_i}{\pi} \ln \frac{1-0.32}{0.32} \right) x_i + 2 \times 0.32 \sum_{i=6}^{10} b_i x_i - \\ \quad \left(2 \times 0.32 - 1\right) \sum_{i=6}^{10} a_i x_i \geq 0.01 - 0.04 \\ \quad \cdots \\ \quad \sum_{i=1}^{5} \left(e_i - \frac{\sqrt{6}\sigma_i}{\pi} \ln \frac{1-0.01}{0.01} \right) x_i + 2 \times 0.01 \sum_{i=6}^{10} b_i x_i - \\ \quad \left(2 \times 0.01 - 1\right) \sum_{i=6}^{10} a_i x_i \geq 0.01 - 0.3 \\ \quad x_1 + x_2 + \cdots + x_{10} = 1 \\ \quad x_i \geq 0, \quad i = 1, 2, \cdots, 10. \end{cases} \tag{3.41}$$

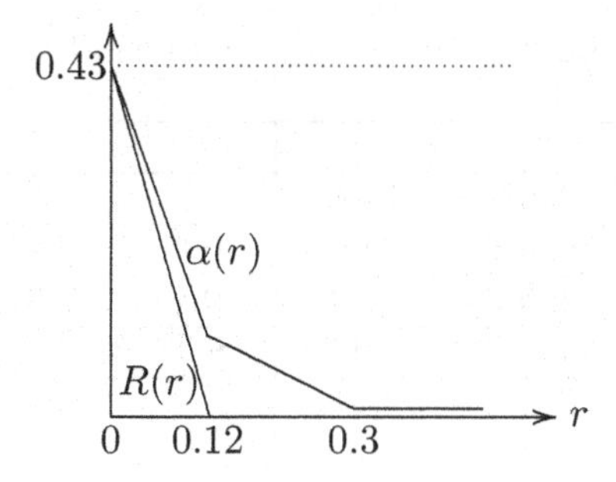

Fig. 3.13 Risk curve $R(r)$ and confidence curve $\alpha(r)$ of Model (3.39).

3.3 β-Return-Risk Model

3.3.1 β-Return-Risk Model

In the mean-risk model, loss is instinct. However, target return is not instinct enough because it is represented by the expected value. Sometimes the investors would like to directly pursue a specific target return rather than an average value. Since the optimal target return may not be obtained in some bad situations, it is natural that people would accept the inability to reach the objective to some extent. However, at a given confidence level which is considered as the safety margin, the objective must be achieved. Based on this idea, Huang [27] proposed the β-return optimization model pursuing the maximal target return at the credibility not less than a predetermined safety level. Replacing expected value by a specific β-return, we get the β-return-risk Model. To understand the β-return-risk selection idea, let us give the definition of β-return and see an example first.

Definition 3.15 *Let x_i be the investment proportions in the i-th securities, $i = 1, 2, \cdots, n$, ξ_i the returns of the i-th securities and β the preset confidence level. The β-return is defined as*

$$\max\{\bar{f} \mid \mathrm{Cr}\left\{\xi_1 x_1 + \xi_2 x_2 + \cdots + \xi_n x_n \geq \bar{f}\right\} \geq \beta\} \tag{3.42}$$

which means the maximal investment return the investor can obtain at confidence level β.

Example 3.39. Suppose we have three securities $\xi_1 = (-0.01, 0.05, 0.1)$, $\xi_2 = (-0.01, 0.06, 0.08)$ and $\xi_3 = (-0.02, 0.08, 0.12)$. There are four money allocation plans. In plan 1, the investor allocates all the money to security 1. In plan 2, the investor allocates all the money to security 2. In plan 3, the investor allocates all the money to security 3. In plan 4, the investor allocates 20% of the money to security 1 and 80% of the money to security 2. The investor sets the confidence level at $\beta = 0.9$. It can be calculated that for plans

Table 3.3 Four Money Allocation Plans

Security i	1	2	3	90%-Return
Money Allocation (plan 1)	100%	0%	0%	0.2%
Money Allocation (plan 2)	0%	100%	0%	0.4%
Money Allocation (plan 3)	0%	0%	100%	0%
Money Allocation (plan 4)	20%	80%	0%	0.36%

1, 2, 3 and 4, the 0.9-return values are $0.2\%, 0.4\%, 0\%, 0.36\%$, respectively. The result is shown in Table 3.3. It can be seen that different money allocation will result in different 0.9-returns. The investor's objective is to find an optimal portfolio which can bring the investor a maximum specific return at a given confidence level, i.e., a maximum β-return. However, a portfolio with maximum β-return may be a risky portfolio. Therefore, before pursuing maximum β-return, the investor has to make sure that the selected portfolio is a safe portfolio. That is to say, the risk curve of the portfolio is first required to be totally below the investor's confidence curve. Then, among the safe portfolios, β-return should be maximized. The mathematical expression of the β-return-risk selection idea is as follows:

$$\begin{cases} \max \bar{f} \\ \text{subject to:} \\ \qquad \mathrm{Cr}\left\{\xi_1 x_1 + \xi_2 x_2 + \cdots + \xi_n x_n \geq \bar{f}\right\} \geq \beta \\ \qquad R(x_1, x_2, \cdots, x_n; r) \leq \alpha(r), \forall r \geq 0 \\ \qquad x_1 + x_2 + \cdots + x_n = 1 \\ \qquad x_i \geq 0, \quad i = 1, 2, \cdots, n \end{cases} \tag{3.43}$$

where $R(x_1, x_2, \cdots, x_n; r)$ is the risk curve of the portfolio, $\alpha(r)$ the investor's confidence curve, and $\bar{f}$ the β-return.

3.3.2 Crisp Equivalent

When security returns are independent fuzzy variables, we can change the β-return-risk model into its equivalent and solve the model in traditional ways.

Theorem 3.14 *Let Φ_i denote the credibility distributions of the i-th fuzzy security return rates $\xi_i, i = 1, 2, \cdots, n$, respectively. Then the β-return-risk model (3.43) can be transformed into the following linear model:*

$$\begin{cases} \max x_1\xi_1(\beta) + x_2\xi_2(\beta) + \cdots + x_n\xi_n(\beta) \\ \textit{subject to:} \\ \qquad x_1\Phi_1^{-1}\Big(\alpha(r)\Big) + x_2\Phi_2^{-1}\Big(\alpha(r)\Big) + \cdots + x_n\Phi_n^{-1}\Big(\alpha(r)\Big) \geq r_f - r, \forall r \geq 0 \\ \qquad x_1 + x_2 + \cdots + x_n = 1 \\ \qquad x_i \geq 0, \quad i = 1, 2, \cdots, n \end{cases} \tag{3.44}$$

where $\xi_i(\beta)$ is the β-return value of the i-th security.

Proof: The objective function follows directly from Theorem 3.11, and the constraint follows from Theorem 3.12 and monotonicity property of credibility measure.

Example 3.40. When all the security returns are regarded to be triangular fuzzy variables $\xi_i = (a_i, b_i, c_i)$, the β-return-risk model (3.43) becomes

$$\begin{cases} \max(2\beta - 1)\sum_{i=1}^{n} a_i x_i + 2(1-\beta)\sum_{i=1}^{n} b_i x_i \\ \text{subject to:} \\ \qquad \Big(2\alpha(r) - 1\Big)\sum_{i=1}^{n} c_i x_i + \Big(2 - 2\alpha(r)\Big)\sum_{i=1}^{n} b_i x_i \geq r_f - r, \text{ if } \alpha(r) > 0.5 \\ \qquad 2\alpha(r)\sum_{i=1}^{n} b_i x_i - \Big(2\alpha(r) - 1\Big)\sum_{i=1}^{n} a_i x_i \geq r_f - r, \text{ if } \alpha(r) \leq 0.5 \\ \qquad x_1 + x_2 + \cdots + x_n = 1 \\ \qquad x_i \geq 0, \quad i = 1, 2, \cdots, n. \end{cases} \tag{3.45}$$

Please note that the objective function is $(2\beta - 1)\sum_{i=1}^{n} a_i x_i + 2(1-\beta)\sum_{i=1}^{n} b_i x_i$ because the confidence level β should be high enough to be greater than 0.5.

Example 3.41. When all the security returns are regarded to be normal fuzzy variables $\xi_i \sim \mathcal{N}(e_i, \sigma_i)$, the β-return-risk model (3.43) becomes

$$\begin{cases} \max \sum_{i=1}^{n} e_i x_i - \frac{\sqrt{6}}{\pi} \ln \frac{\beta}{1-\beta} \sum_{i=1}^{n} \sigma_i x_i \\ \text{subject to:} \\ \qquad \sum_{i=1}^{n} \left(e_i - \frac{\sqrt{6}\sigma_i}{\pi} \ln \frac{1-\alpha(r)}{\alpha(r)} \right) x_i \geq r_f - r, \quad \forall r \geq 0 \\ \qquad x_1 + x_2 + \cdots + x_n = 1 \\ \qquad x_i \geq 0, \quad i = 1, 2, \cdots, n. \end{cases} \tag{3.46}$$

The objective function is $\sum_{i=1}^{n} e_i x_i - \frac{\sqrt{6}}{\pi} \ln \frac{\beta}{1-\beta} \sum_{i=1}^{n} \sigma_i x_i$ because the confidence level β should be high enough to be greater than 0.5.

Example 3.42. Suppose the return rates of the i-th securities are normal fuzzy variables $\xi_i \sim \mathcal{N}(e_i, \sigma_i), i = 1, 2, \cdots, m$, and the return rates of the j-th securities are triangular fuzzy variables $\xi_j = (a_j, b_j, c_j), j = m+1, m+2, \cdots, n$, respectively. Then the fuzzy β-return-risk model can be transformed into the following form:

$$\begin{cases} \max \sum_{i=1}^{m} e_i x_i - \frac{\sqrt{6}}{\pi} \ln \frac{\beta}{1-\beta} \sum_{i=1}^{m} \sigma_i x_i + \\ \sum_{i=m+1}^{n} (2\beta - 1) a_i x_i + 2(1-\beta) \sum_{i=m+1}^{n} b_i x_i \\ \text{subject to:} \\ \sum_{i=1}^{m} \left(e_i - \frac{\sqrt{6}\sigma_i}{\pi} \ln \frac{1-\alpha(r)}{\alpha(r)} \right) x_i + \left(2\alpha(r) - 1\right) \sum_{i=m+1}^{n} c_i x_i + \\ \left(2 - 2\alpha(r)\right) \sum_{i=m+1}^{n} b_i x_i \geq r_f - r, \text{ if } \alpha(r) > 0.5 \\ \sum_{i=1}^{m} \left(e_i - \frac{\sqrt{6}\sigma_i}{\pi} \ln \frac{1-\alpha(r)}{\alpha(r)} \right) x_i + 2\alpha(r) \sum_{i=m+1}^{n} b_i x_i - \\ \left(2\alpha(r) - 1\right) \sum_{i=m+1}^{n} a_i x_i \geq r_f - r, \text{ if } \alpha(r) \leq 0.5 \\ x_1 + x_2 + \cdots + x_n = 1 \\ x_i \geq 0, \quad i = 1, 2, \cdots, n. \end{cases} \tag{3.47}$$

Since the sum weighted normal fuzzy variable is still a normal fuzzy variable, $\sum_{i=1}^{m} \xi_i x_i$ is a normal fuzzy variable $\mathcal{N}\left(\sum_{i=1}^{m} e_i x_i, \sum_{i=1}^{m} \sigma_i x_i\right)$. Since the sum weighted triangular fuzzy variable is still a triangular fuzzy variable, $\sum_{i=m+1}^{n} \xi_i x_i$ is a triangular fuzzy variable $\left(\sum_{i=m+1}^{n} a_i x_i, \sum_{i=m+1}^{n} b_i x_i, \sum_{i=m+1}^{n} c_i x_i \right)$. Since the confidence level β should be high enough to be greater than 0.5, we can get Model (3.47) from Models (3.45) and (3.46) directly.

3.3.3 An Example

Suppose an investor wants to choose an optimal portfolio from ten securities whose return rates are given in Table 3.1 in Subsection 3.2.5. The monthly risk-free interest rate is still 0.01, and the investor gives his/her confidence curve is the same as follows:

$$\alpha(r) = \begin{cases} -2.75r + 0.43, & 0 \le r \le 0.12, \\ -0.5r + 0.16, & 0.12 \le r \le 0.3, \\ 0.01, & r \ge 0.3. \end{cases}$$

Suppose this time, the investor wants to pursue a maximum return value at confidence 0.95 among the safe portfolios. According to the β-return-risk selection idea, we build the model as follows:

$$\begin{cases} \max \bar{f} \\ \text{subject to:} \\ \qquad \mathrm{Cr}\{\xi_1 x_1 + \xi_2 x_2 + \cdots + \xi_{10} x_{10} \ge \bar{f}\} \ge 0.95 \\ \qquad R(x_1, x_2, \cdots, x_{10}; r) \le \alpha(r), \ \forall r \ge 0 \\ \qquad x_1 + x_2 + \cdots + x_{10} = 1 \\ \qquad x_i \ge 0, \quad i = 1, 2, \cdots, 10 \end{cases} \tag{3.48}$$

where $R(x_1, x_2, \cdots, x_{10}; r)$ is the risk curve of the portfolio defined as

$$R(x_1, x_2, \cdots, x_{10}; r) = \mathrm{Cr}\{0.01 - (\xi_1 x_1 + \xi_2 x_2 + \cdots + \xi_{10} x_{10}) \ge r\}.$$

According to Model (3.47), we can change Model (3.48) into the following linear programming form. Note that $\alpha(r) < 0.5$ in the example.

$$\begin{cases} \max \sum_{i=1}^{5} e_i x_i - \frac{\sqrt{6}}{\pi} \sum_{i=1}^{5} \sigma_i x_i \ln 19 + 0.9 \sum_{i=6}^{10} a_i x_i + 0.1 \sum_{i=6}^{10} b_i x_i \\ \text{subject to:} \\ \qquad \sum_{i=1}^{5} \left(e_i - \frac{\sqrt{6}\sigma_i}{\pi} \ln \frac{1 - \alpha(r)}{\alpha(r)} \right) x_i + 2\alpha(r) \sum_{i=6}^{10} b_i x_i - \\ \qquad \left(2\alpha(r) - 1 \right) \sum_{i=6}^{10} a_i x_i \ge r_f - r \\ \qquad x_1 + x_2 + \cdots + x_{10} = 1 \\ \qquad x_i \ge 0, \quad i = 1, 2, \cdots, 10. \end{cases} \tag{3.49}$$

Since the investor's confidence curve is a horizontal line when $r \ge 0.3$ and risk curve of the portfolio is a decreasing function of r, when checking if risk curve of the portfolio is totally below the investor's confidence curve, it is enough to check if the points on the risk curve are all lower than the points on the confidence curve for $(r = 0, \alpha = 0.43)$, $(r = 0.02, \alpha = 0.375)$, $(r =$

Table 3.4 Allocation of Money to Ten Securities

Security i	1	2	3	4	5
Allocation of money	0.00%	0.00 %	0.00%	0.00%	0.00%
Security i	6	7	8	9	10
Allocation of money	60%	40 %	0.00%	0.00%	0.00%

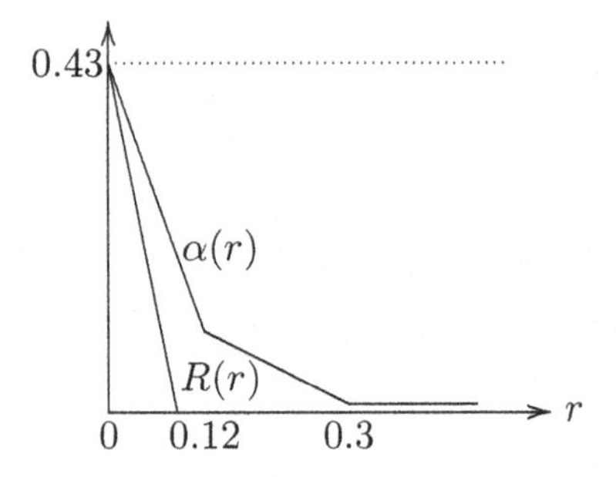

Fig. 3.14 Risk curve $R(r)$ and confidence curve $\alpha(r)$ of Model (3.49).

$0.04, \alpha = 0.32)$, $(r = 0.06, \alpha = 0.265), \cdots, (r = 0.3, \alpha = 0.01)$. By using "Solver" in "Excel", we get the optimal portfolio shown in Table 3.4. The maximum return the investor can obtain at credibility 0.95 is -0.066. As shown in Fig. 3.14, risk curve of the optimal portfolio is totally below the investor's confidence curve. Given every loss value r, the loss occurrence credibility is not greater than the investor's tolerable credibility. Or given every occurrence

Table 3.5 Optimal Portfolios Produced by Different Selection Criteria

Optimal Portfolio	Mean-Risk Criterion	β-Return-Risk Criterion
ξ_1	0.00%	0.00%
ξ_2	0.00 %	0.00%
ξ_3	0.00%	0.00%
ξ_4	0.00%	86.02%
ξ_5	0.00 %	13.98%
ξ_6	0.00 %	0.00%
ξ_7	78.57%	0.00%
ξ_8	0.00%	0.00%
ξ_9	0.00 %	0.00%
ξ_{10}	21.43 %	0.00%
Expected Return	4.2%	3.55%
0.95-Return	−9.9%	−6.6%

credibility $\alpha(r)$, the loss level is not greater than the investor's tolerable loss level.

Remark 3.7. We put the results of examples of mean-risk model in subsection 3.2.5 and β-return-risk model in this subsection together in Table 3.5. It is seen that even when risk-free interest rate, the alternative individual securities and the investor's confidence curve are same, adopting different selection criteria produces different results.

3.4 Credibility Minimization Model

3.4.1 Credibility Minimization Model

When investors are sensitive to only one disastrous low return level, risk curve will degenerate to the credibility of a portfolio return below the specific disaster level, which is proposed to be an alternative definition of risk by Huang [27] in fuzzy portfolio selection. If the investors adopt this definition of risk, they will select the portfolio with minimum occurrence credibility of the specific disastrous return level. Let ξ_i be the i-th security returns and x_i the investment proportions, $i = 1, 2, \cdots, n$, respectively. Taking investment return into account, the investors should select the portfolio whose expected return is not less than a preset expected value and in the meantime whose occurrence credibility of the sensitive bad event is minimum. The selection idea of minimizing the occurrence credibility of the sensitive bad event is expressed as follows:

$$\begin{cases} \min \mathrm{Cr}\left\{\xi_1 x_1 + \xi_2 x_2 + \cdots + \xi_n x_n \leq d\right\} \\ \text{subject to:} \\ \qquad E[\xi_1 x_1 + \xi_2 x_2 + \cdots + \xi_n x_n] \geq a \\ \qquad x_1 + x_2 + \cdots + x_n = 1 \\ \qquad x_i \geq 0, \quad i = 1, 2, \cdots, n \end{cases} \tag{3.50}$$

where d is the concerned disastrous low return level and a the preset minimum expected return that the investors can accept.

Let us recall the definition of risk curve. The curve

$$R(x_1, x_2, \cdots, x_n; r) = \mathrm{Cr}\{r_f - (\xi_1 x_1 + \xi_2 x_2 + \cdots + \xi_n x_n) \geq r\}, \forall r \geq 0$$

is called the risk curve of the portfolio, where r_f is the risk-free interest rate. Let r degenerate to one specific number r_0, then the risk curve becomes

$$\begin{aligned} R(x_1, x_2, \cdots, x_n; r_0) &= \mathrm{Cr}\{r_f - (\xi_1 x_1 + \xi_2 x_2 + \cdots + \xi_n x_n) \geq r_0\} \\ &= \mathrm{Cr}\{\xi_1 x_1 + \xi_2 x_2 + \cdots + \xi_n x_n \leq r_f - r_0\} \end{aligned}$$

which is just the risk definition of occurrence credibility of a sensitive low return event. It is clear that $r_f - r_0 = d$.

If the investors pre-give a confidence level α, what will be the maximum potential loss for the given confidence level? We can use Value-at-Risk-in-Fuzziness (VaRF) to answer the question.

Definition 3.16 *Let ξ denote a fuzzy return of a portfolio, and r_f the risk-free interest rate. Then Value-at-Risk-in-Fuzziness* (VaRF) *is defined as*

$$\mathrm{VaRF}(\alpha) = \sup\{\bar{r} | \mathrm{Cr}\{r_f - \xi \geq \bar{r}\} \leq 1 - \alpha\}. \tag{3.51}$$

where α is the preset confidence level.

For example, if VaRF(95%) = 8%, it means that there is only a 5% credibility that the portfolio return rate will drop more than 8% below the risk-free interest rate. It is easy to see that VaRF is in fact an inverse version of the risk definition of the credibility of a portfolio return below a specific disaster level.

If the investors adopt VaRF as the investment risk, they will select the portfolio with minimum VaRF value. Taking investment return into account, the investors should select the portfolio whose expected return is not less than a preset level and in the meantime whose VaRF value is the minimum. The selection idea of minimizing VaRF can be expressed by the following model:

$$\begin{cases} \min \bar{r} \\ \text{subject to:} \\ \qquad \mathrm{Cr}\left\{r_f - (\xi_1 x_1 + \xi_2 x_2 + \cdots + \xi_n x_n) \geq \bar{r}\right\} \leq 1 - \alpha \\ \qquad E[\xi_1 x_1 + \xi_2 x_2 + \cdots + \xi_n x_n] \geq a \\ \qquad x_1 + x_2 + \cdots + x_n = 1 \\ \qquad x_i \geq 0, \quad i = 1, 2, \cdots, n \end{cases} \tag{3.52}$$

where a is the pre-set tolerable minimum expected return, α the pre-determined confidence level, and $\bar{r}$ the VaRF defined as

$$\mathrm{VaRF}(\alpha) = \sup\{\bar{r} | \mathrm{Cr}\{r_f - (\xi_1 x_1 + \xi_2 x_2 + \cdots + \xi_n x_n) \geq \bar{r}\} \leq 1 - \alpha\}.$$

It is seen that the VaRF minimization model (3.52) can be regarded as another version of credibility minimization model (3.50).

Mathematically, Model (3.52) is a minmax model because it is equivalent to

$$\begin{cases} \min\limits_{x_1,x_2,\cdots,x_n} \max\limits_{\bar{r}} \bar{r} \\ \text{subject to:} \\ \qquad \mathrm{Cr}\{r_f - (\xi_1 x_1 + \xi_2 x_2 + \cdots + \xi_n x_n) \geq \bar{r}\} \leq 1-\alpha \\ \qquad E[\xi_1 x_1 + \xi_2 x_2 + \cdots + \xi_n x_n] \geq a \\ \qquad x_1 + x_2 + \cdots + x_n = 1 \\ \qquad x_i \geq 0, \quad i = 1, 2, \cdots, n \end{cases} \tag{3.53}$$

where $\max \bar{r}$ is the VaRF.

3.4.2 Crisp Equivalent

In some special cases, we can convert the credibility minimization model (3.50) into its crisp equivalent.

Example 3.43. When people invest, the credibility of portfolio return equal to or lower than a sensitive disaster level d should always be required to be less than 0.5. Thus, when security returns are regarded to be all triangular fuzzy variables $\xi_i = (a_i, b_i, c_i)$, the credibility minimization model (3.50) becomes

$$\begin{cases} \min \left(d - \sum\limits_{i=1}^{n} a_i x_i\right) \Big/ \left(\sum\limits_{i=1}^{n} b_i x_i - \sum\limits_{i=1}^{n} a_i x_i\right) \\ \text{subject to:} \\ \qquad \sum\limits_{i=1}^{n} a_i x_i \leq d \leq \sum\limits_{i=1}^{n} b_i x_i \\ \qquad \sum\limits_{i=1}^{n} (a_i x_i + 2 b_i x_i + c_i x_i) \geq 4a \\ \qquad x_1 + x_2 + \cdots + x_n = 1 \\ \qquad x_i \geq 0, \quad i = 1, 2, \cdots, n. \end{cases} \tag{3.54}$$

Please note that the constraint $\sum\limits_{i=1}^{n} a_i x_i \leq d \leq \sum\limits_{i=1}^{n} b_i x_i$ is added because the credibility of portfolio return equal to or lower than the concerned disaster level d should be less than 0.5.

Example 3.44. When security returns are regarded to be all normal fuzzy variables $\xi_i \sim \mathcal{N}(e_i, \sigma_i)$, the credibility minimization model (3.50) becomes

$$\begin{cases} \max \left(\sum_{i=1}^{n} e_i x_i - d\right) \Big/ \sum_{i=1}^{n} \sigma_i x_i \\ \text{subject to:} \\ \qquad d \le e_1 x_1 + e_2 x_2 + \cdots + e_n x_n \\ \qquad e_1 x_1 + e_2 x_2 + \cdots + e_n x_n \ge a \\ \qquad x_1 + x_2 + \cdots + x_n = 1 \\ \qquad x_i \ge 0, \quad i = 1, 2, \cdots, n. \end{cases} \tag{3.55}$$

Model (3.55) can easily be obtained because the sum of weighted normal fuzzy variables is still a normal fuzzy variable. Please note that to minimize the credibility value

$$\left(1 + \exp\left(\pi\Big(\sum_{i=1}^{n} e_i x_i - d\Big) \Big/ \sqrt{6} \sum_{i=1}^{n} \sigma_i x_i\right)\right)^{-1}$$

we just need to maximize

$$\Big(\sum_{i=1}^{n} e_i x_i - d\Big) \Big/ \sum_{i=1}^{n} \sigma_i x_i.$$

A constraint $d \le e_1 x_1 + e_2 x_2 + \cdots + e_n x_n$ is added because the chance of portfolio return equal to or less than the concerned disaster level d should be less than 0.5.

Theorem 3.15 *Let Φ_i denote the credibility distributions of the i-th fuzzy security return rates $\xi_i, i = 1, 2, \cdots, n$, respectively. Then the* VaRF *minimization model (3.52) can be transformed into the following linear model:*

$$\begin{cases} \min r_f - x_1 \Phi_1^{-1}(1-\alpha) - x_2 \Phi_2^{-1}(1-\alpha) + \cdots - x_n \Phi_n^{-1}(1-\alpha) \\ \textit{subject to:} \\ \qquad x_1 E[\xi_1] + x_2 E[\xi_2] + \cdots + x_n E[\xi_n] \ge a \\ \qquad x_1 + x_2 + \cdots + x_n = 1 \\ \qquad x_i \ge 0, \quad i = 1, 2, \cdots, n. \end{cases} \tag{3.56}$$

Proof: It follows directly from Theorem 3.12 and the monotonicity property of credibility measure.

Example 3.45. Suppose the return rates of the i-th securities are normal fuzzy variables $\xi_i \sim \mathcal{N}(e_i, \sigma_i), i = 1, 2, \cdots, m$, and the return rates of the j-th securities are triangular fuzzy variables $\xi_j = (a_j, b_j, c_j), j = m+1, m+2, \cdots, n$, respectively. Since confidence level $\alpha > 0.5$ and $1-\alpha < 0.5$, the VaRF minimization model (3.52) can be transformed into the following form:

$$
\begin{cases}
\min r_f - \sum_{i=1}^{m}\left(e_i - \frac{\sqrt{6}\sigma_i}{\pi}\ln\frac{\alpha}{1-\alpha}\right)x_i - \\
\qquad 2(1-\alpha)\sum_{i=m+1}^{n} b_i x_i + \left(1-2\alpha\right)\sum_{i=m+1}^{n} a_i x_i \\
\text{subject to:} \\
\qquad \sum_{i=1}^{m} e_i x_i + \sum_{i=m+1}^{n}\frac{1}{4}(a_i x_i + 2b_i x_i + c_i x_i) \geq a \\
\qquad x_1 + x_2 + \cdots + x_n = 1 \\
\qquad x_i \geq 0, \quad i = 1, 2, \cdots, n.
\end{cases}
\tag{3.57}
$$

3.4.3 An Example

Suppose an investor wants to choose an optimal portfolio from ten securities of which five security return rates are normal fuzzy variables and the rest five the triangular fuzzy variables. The prediction of the return rates of the ten securities is given in Table 3.6. The risk-free interest rate is assumed to be 0.01. Suppose the minimum expected return the investor can accept is 0.03, and the investor wants to minimize the specific potential loss at confidence level 0.95. Then according to the VaRF minimization selection idea , we build the model as follows:

$$
\begin{cases}
\min \bar{r} \\
\text{subject to:} \\
\qquad \mathrm{Cr}\{0.01 - (\xi_1 x_1 + \xi_2 x_2 + \cdots + \xi_{10} x_{10}) \geq \bar{r}\} \leq 0.05 \\
\qquad E[\xi_1 x_1 + \xi_2 x_2 + \cdots + \xi_{10} x_{10}] \geq 0.03 \\
\qquad x_1 + x_2 + \cdots + x_{10} = 1 \\
\qquad x_i \geq 0, \quad i = 1, 2, \cdots, 10.
\end{cases}
\tag{3.58}
$$

According to model (3.57), we change the model (3.58) into the following form:

$$
\begin{cases}
\min 0.01 - \sum_{i=1}^{5}\left(e_i - \frac{\sqrt{6}\sigma_i}{\pi}\ln 19\right)x_i - \sum_{i=6}^{10}(0.1b_i x_i + 0.9a_i x_i) \\
\text{subject to:} \\
\qquad \sum_{i=1}^{5} e_i x_i + \sum_{i=6}^{10}\frac{1}{4}(a_i x_i + 2b_i x_i + c_i x_i) \geq 0.03 \\
\qquad x_1 + x_2 + \cdots + x_{10} = 1 \\
\qquad x_i \geq 0, \quad i = 1, 2, \cdots, 10.
\end{cases}
\tag{3.59}
$$

Table 3.6 Fuzzy Return Rates of Ten Securities

Security i	$\xi_i \sim \mathcal{N}(e_i, \sigma_i)$	Security i	$\xi_i = (a_i, b_i, c_i)$
1	$\mathcal{N}(0.033, 0.44)$	6	$(-0.008, 0.026, 0.06)$
2	$\mathcal{N}(0.032, 0.40)$	7	$(-0.02, 0.030, 0.08)$
3	$\mathcal{N}(0.039, 0.45)$	8	$(-0.01, 0.032, 0.08)$
4	$\mathcal{N}(0.031, 0.39)$	9	$(-0.05, 0.04, 0.10)$
5	$\mathcal{N}(0.025, 0.32)$	10	$(-0.03, 0.03, 0.09)$

Table 3.7 Allocation of Money to Ten Securities

Security i	1	2	3	4	5
Allocation of money	0.00%	0.00 %	0.00%	0.00%	0.00%
Security i	6	7	8	9	10
Allocation of money	46.67%	0.00%	53.33%	0.00%	0.00%

Using "Solver" in "Excel", we obtain the optimal portfolio shown in Table 3.7. The minimum objective value is 0.015, which means that if the investor invests 46.67% of his/her money in security 6 and 53.33% in security 8, the expected return will not be lower than 0.03, and in the meantime there is only a 5% credibility that the portfolio return rate will drop more than 1.5% below the risk-free interest rate.

3.5 Mean-Variance Model

3.5.1 Mean-Variance Model

Risk curve takes a panoramic view of the whole likely loss events. Sometimes people wish to use average information to evaluate the risk. As a counterpart of Markowitz's mean-variance model, Huang [33] proposed credibilistic mean-variance model for portfolio selection with fuzzy returns.

Let ξ_i represent the fuzzy returns of the i-th securities and x_i the investment proportions in the i-th securities, $i = 1, 2, \cdots, n$, respectively. The philosophy of the mean-variance model is to pursue the maximum expected return with the variance not greater than the preset level. Let γ be the maximum variance level the investors can tolerate. The credibilistic mean-variance selection model is expressed as follows:

$$\begin{cases} \max E[x_1\xi_1 + x_2\xi_2 + \cdots + x_n\xi_n] \\ \text{subject to:} \\ \qquad V[x_1\xi_1 + x_2\xi_2 + \cdots + x_n\xi_n] \le \gamma \\ \qquad x_1 + x_2 + \cdots + x_n = 1 \\ \qquad x_i \ge 0, \quad i = 1, 2, \cdots, n \end{cases} \tag{3.60}$$

where E denotes the expected value operator, and V the variance operator of the fuzzy variables. The constraint $V[x_1\xi_1+x_2\xi_2+\cdots+x_n\xi_n]\le\gamma$ ensures that the optimal portfolio will be selected only from the portfolios whose average square deviations from the expected return are not greater than the tolerable level.

Sometimes, the investors may preset a level of expected return. Then the philosophy of mean-variance model becomes to minimize variance value of the portfolio with the expected value of the portfolio not less than this preset expected return level. Thus the credibilistic mean-variance model is expressed in the following way:

$$\begin{cases} \min V[x_1\xi_1+x_2\xi_2+\cdots+x_n\xi_n] \\ \text{subject to:} \\ \qquad E[x_1\xi_1+x_2\xi_2+\cdots+x_n\xi_n]\ge\lambda \\ \qquad x_1+x_2+\cdots+x_n=1 \\ \qquad x_i\ge 0,\quad i=1,2,\cdots,n \end{cases} \tag{3.61}$$

where λ represents the minimum expected return the investors feel satisfactory. The constraint $E[x_1\xi_1+x_2\xi_2+\cdots+x_n\xi_n]\ge\lambda$ ensures that the optimal portfolio is selected only among those satisfactory portfolios, i.e., the portfolios whose expected return will not be less than the preset expected return level.

From Models (3.60) and (3.61), we can see that if we change the preset variance value or expected value, we will get different optimal solution. A portfolio is efficient if it is impossible to obtain higher expected return with no greater variance value, or it is impossible to obtain less variance value with no less expected return. All efficient portfolios make up the efficient frontier. An efficient portfolio is in fact an solution of the following optimization model with two objectives:

$$\begin{cases} \max E[x_1\xi_1+x_2\xi_2+\cdots+x_n\xi_n] \\ \min V[x_1\xi_1+x_2\xi_2+\cdots+x_n\xi_n] \\ \text{subject to:} \\ \qquad x_1+x_2+\cdots+x_n=1 \\ \qquad x_i\ge 0,\quad i=1,2,\cdots,n. \end{cases} \tag{3.62}$$

Different investors will find different optimal portfolios from the efficient frontier according to their own preferences to risk aversion, i.e., tradeoff of variance and expected return.

3.5.2 Crisp Equivalent

According to the properties of triangular fuzzy variable, trapezoidal fuzzy variable and normal fuzzy variable, we give the crisp equivalents of credibilistic mean-variance model in some special cases.

When all the security returns are described by symmetrical triangular fuzzy variables $\xi_i = (a_i, b_i, c_i)$, the mean-variance model (3.60) becomes

$$\begin{cases} \max \sum_{i=1}^{n} a_i x_i + 2\sum_{i=1}^{n} b_i x_i + \sum_{i=1}^{n} c_i x_i \\ \text{subject to:} \\ \qquad \sum_{i=1}^{n} c_i x_i - \sum_{i=1}^{n} a_i x_i \le \sqrt{24\gamma} \\ \qquad x_1 + x_2 + \cdots + x_n = 1 \\ \qquad x_i \ge 0, \quad i = 1, 2, \cdots, n. \end{cases} \tag{3.63}$$

When all the security returns are described by symmetrical trapezoidal fuzzy variables $\xi_i = (a_i, b_i, c_i, d_i)$, the mean-variance model (3.60) becomes

$$\begin{cases} \max \sum_{i=1}^{n} a_i x_i + \sum_{i=1}^{n} b_i x_i + \sum_{i=1}^{n} c_i x_i + \sum_{i=1}^{n} d_i x_i \\ \text{subject to:} \\ \qquad \left(\sum_{i=1}^{n} d_i x_i - \sum_{i=1}^{n} a_i x_i\right)^2 + \left(\sum_{i=1}^{n} c_i x_i - \sum_{i=1}^{n} b_i x_i\right)^2 \\ \qquad +\left(\sum_{i=1}^{n} d_i x_i - \sum_{i=1}^{n} a_i x_i\right)\left(\sum_{i=1}^{n} c_i x_i - \sum_{i=1}^{n} b_i x_i\right) \le 24\gamma \\ \qquad x_1 + x_2 + \cdots + x_n = 1 \\ \qquad x_i \ge 0, \quad i = 1, 2, \cdots, n. \end{cases} \tag{3.64}$$

When all the security returns are described by normal fuzzy variables $\xi_i \sim \mathcal{N}(e_i, \sigma_i)$, the mean-variance model (3.60) becomes

$$\begin{cases} \max e_1 x_1 + e_2 x_2 + \cdots + e_n x_n \\ \text{subject to:} \\ \qquad \sigma_1 x_1 + \sigma_2 x_2 + \cdots + \sigma_n x_n \le \sqrt{\gamma} \\ \qquad x_1 + x_2 + \cdots + x_n = 1 \\ \qquad x_i \ge 0, \quad i = 1, 2, \cdots, n. \end{cases} \tag{3.65}$$

3.5.3 An Example

Suppose an investor wants to select his/her portfolio from the ten securities whose returns are given in Table 3.8. If the investor adopts the mean-variance selection idea, and set the minimum expected return at 0.07. Then according to the credibilistic mean-variance model, the investor should select the portfolio according to the following model:

$$\begin{cases} \min V[\xi_1 x_1 + \xi_2 x_2 + \cdots + \xi_{10} x_{10}] \\ \text{subject to:} \\ \qquad E[\xi_1 x_1 + \xi_2 x_2 + \cdots + \xi_{10} x_{10}] \geq 0.07 \\ \qquad x_1 + x_2 + \cdots + x_{10} = 1 \\ \qquad x_i \geq 0, \quad i = 1, 2, \cdots, 10. \end{cases} \tag{3.66}$$

According to the properties of triangular fuzzy variable, we change Model (3.66) into the following crisp form:

$$\begin{cases} \min \sum_{i=1}^{10} c_i x_i - \sum_{i=1}^{10} a_i x_i \\ \text{subject to:} \\ \qquad \sum_{i=1}^{10} (a_i x_i + 2b_i x_i + c_i x_i) \geq 0.28 \\ \qquad x_1 + x_2 + \cdots + x_{10} = 1 \\ \qquad x_i \geq 0, \quad i = 1, 2, \cdots, 10. \end{cases} \tag{3.67}$$

Table 3.8 Fuzzy Return Rates of Ten Securities

Security i	$\xi_i = (a_i, b_i, c_i)$	Security i	$\xi_i = (a_i, b_i, c_i)$
1	(-0.08, 0.02, 0.12)	6	$(-0.09, 0.06, 0.15)$
2	(-0.1, 0.04, 0.14)	7	$(-0.16, 0.09, 0.25)$
3	(-0.12, 0.05, 0.17)	8	$(-0.18, 0.06, 0.24)$
4	(-0.11, 0.06, 0.17)	9	$(-0.15, 0.08, 0.23)$
5	(-0.12, 0.06, 0.18)	10	$(-0.22, 0.1, 0.32)$

Table 3.9 Allocation of Money to Ten Securities

Security i	1	2	3	4	5
Allocation of money	0.00%	0.00 %	0.00%	0.00%	0.00%
Security i	6	7	8	9	10
Allocation of money	0.00 %	66.67 %	0.00 %	0.00%	33.33%

By using "Solver" in "Excel", in order to minimize the variance with the expected return not less than 0.07, the investor should allocate his/her money according to Table 3.9. The objective value of Model (3.67) is 0.433, which means the minimum variance is $0.433^2/24 = 0.0086$.

3.5.4 Mean-Semivariance Model

When the membership functions of the security returns are asymmetrical, variance becomes a deficient measure of risk because when eliminating variance both lower and higher deviations from the expected value are eliminated, yet higher deviations are what we want. Empirical evidences [3, 12, 16, 86] show that there do exist cases that security returns are not symmetrically distributed. Therefore, Huang [37] defined semivariance of fuzzy variable that only measures the lower deviation from the expected value and proposed credibilistic mean-semivariance model in which semivariance replaces variance as the measure of risk.

Let x_i be the investment proportions in securities i, and ξ_i the i-th fuzzy security returns, $i = 1, 2, \cdots, n$, respectively. Similar to mean-variance model, the philosophy of the mean-semivariance model is to maximize the expected return at the given level of risk. Substituting variance with semivariance, we have the credibilistic mean-semivariance model as follows:

$$\begin{cases} \max E[x_1\xi_1 + x_2\xi_2 + \cdots + x_n\xi_n] \\ \text{subject to:} \\ \qquad SV[x_1\xi_1 + x_2\xi_2 + \cdots + x_n\xi_n] \le \gamma \\ \qquad x_1 + x_2 + \cdots + x_n = 1 \\ \qquad x_i \ge 0, \quad i = 1, 2, \cdots, n \end{cases} \tag{3.68}$$

where γ denotes the maximum semivariance level the investors can tolerate, E the expected value operator, and SV the semivariance of the fuzzy variables.

Sometimes the investors may preset a minimum acceptable expected return level, then the mean-semivariance model is expressed as follows:

$$\begin{cases} \min SV[x_1\xi_1 + x_2\xi_2 + \cdots + x_n\xi_n] \\ \text{subject to:} \\ \qquad E[x_1\xi_1 + x_2\xi_2 + \cdots + x_n\xi_n] \ge \lambda \\ \qquad x_1 + x_2 + \cdots + x_n = 1 \\ \qquad x_i \ge 0, \quad i = 1, 2, \cdots, n \end{cases}$$

where λ denotes the minimum expected return that the investors can accept.

From Theorem 3.8 we know that when the membership functions of the security returns are symmetrical, optimal portfolio can be obtained no matter whether we take the variance or the semivariance as the measurement of risk.

However, when membership functions of security returns are asymmetrical, Theorem 3.7 indicates that taking semivariance or variance as the measurement of risk will yield different results. So semivariance can be regarded as an improvement of variance as the measure of risk because semivariance is free from the reliance on symmetrical security returns.

3.6 Entropy Optimization Model

Given a fuzzy portfolio return, investors will usually regard the portfolio to be risky if it is difficult to predict the specific value that the portfolio return may take. Fuzzy entropy measures the difficulty degree of the prediction. When a portfolio return distributes dispersively, the entropy of the return is great, which implies that the return contains much uncertainty and the prediction is difficult; when the portfolio return distributes concentratively, the entropy of the return is small, which implies that the return contains little uncertainty and the prediction is easy. In addition, entropy can well reflect the dispersive degree of the portfolio return no matter if the membership function of the portfolio return is symmetrical or not. Therefore, Huang [39] suggested entropy being another alternative measure of risk and proposed entropy optimization model in which the philosophy is to pursue the maximum expected return among the portfolios whose return distribution is concentrative enough to the required level.

Let ξ_i denote the i-th fuzzy security returns, x_i the investment proportions, $i = 1, 2, \cdots, n$, respectively, and γ the preset entropy level. The mean-entropy model is as follows:

$$\begin{cases} \max E[x_1\xi_1 + x_2\xi_2 + \cdots + x_n\xi_n] \\ \text{subject to:} \\ \qquad H[x_1\xi_1 + x_2\xi_2 + \cdots + x_n\xi_n] \le \gamma \\ \qquad x_1 + x_2 + \cdots + x_n = 1 \\ \qquad x_i \ge 0, \quad i = 1, 2, \cdots, n \end{cases} \tag{3.69}$$

where E is the expected value operator and H the entropy. The constraint $H[x_1\xi_1 + x_2\xi_2 + \cdots + x_n\xi_n] \le \gamma$ means that the optimal portfolio must be selected from the portfolios whose returns are concentrative enough to be less than a preset tolerable level. Compared with the mean-variance model (3.60), entropy is more general than variance as a measure of risk because entropy is free from reliance on symmetrical distribution of the security returns, i.e., entropy remains an effective measure of risk when the membership functions of the security returns are asymmetrical. However, when security returns are symmetrical triangular fuzzy variables or the normally distributed fuzzy variables, the optimal solution of the mean-entropy model (3.69) is also the optimal solution of of the mean-variance model (3.60) and vice versa because in these two special cases, entropies can be expressed by the product of some constant and the corresponding variances.

3.7 Hybrid Intelligent Algorithm

We have given the crisp equivalents of the fuzzy portfolio selection models in some special cases so that we can find the optimal portfolios by traditional methods. However, in many cases it is difficult to convert the fuzzy selection models into their equivalents. To produce a general solution algorithm, we integrate fuzzy simulation and genetic algorithm (GA) to produce the hybrid intelligent algorithm. In our algorithm, generally speaking, fuzzy simulation is used to calculate the objective and constraint values, and GA is employed to find the optimal solution. A scheme of the algorithm (Fig. 3.15) is given as follows:

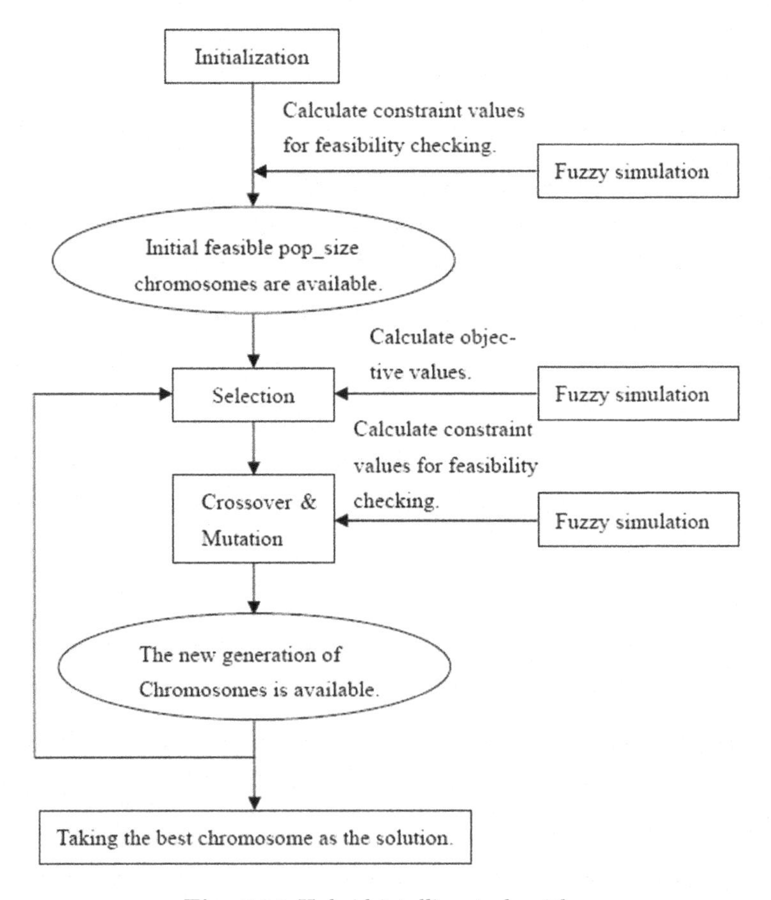

Fig. 3.15 Hybrid intelligent algorithm.

3.7.1 Fuzzy Simulation

Fuzzy simulation has been studied by many scholars. In particular, Liu [56] introduced in detail the technique based on the concept of credibility. Here, we will introduce the simulation procedure for calculating the objective and constraint values appeared in our optimization models, i.e., the expected value, the variance value, the semivariance value, the β-return value, the credibility value, and the entropy value of the fuzzy portfolio return.

Let ξ_i be fuzzy returns with membership functions μ_i, and x_i the investment proportions, $i = 1, 2, \cdots, n$, respectively, where n is the number of securities. For convenience, let $\boldsymbol{\xi} = (\xi_1, \xi_2, \cdots, \xi_n)$, and $\boldsymbol{x} = (x_1, x_2, \cdots, x_n)$. Let $\boldsymbol{\mu} = (\mu_1, \mu_2, \cdots, \mu_n)$, and denote the membership function vector of $\boldsymbol{\xi}$. Since the variance value and the semivariance value is a kind of expected value, we can know the simulation procedure for them if we know the simulation procedure for calculating the expected value. In addition, if we know how to calculate $\mathrm{Cr}\{f(\boldsymbol{x}, \boldsymbol{\xi}) \geq r\}$, we can know how to calculate $\mathrm{Cr}\{f(\boldsymbol{x}, \boldsymbol{\xi}) \leq r\}$ because $\mathrm{Cr}\{f(\boldsymbol{x}, \boldsymbol{\xi}) \leq r\} = \mathrm{Cr}\{-f(\boldsymbol{x}, \boldsymbol{\xi}) \geq -r\} = \mathrm{Cr}\{f^{'}(\boldsymbol{x}, \boldsymbol{\xi}) \geq r^{'}\}$. Thus, we in fact only need to calculate the values of the following four types of uncertain functions:

$$
\begin{aligned}
U_1: &\quad \boldsymbol{x} \to \mathrm{Cr}\{f(\boldsymbol{x}, \boldsymbol{\xi}) \geq r\}, \\
U_2: &\quad \boldsymbol{x} \to E\left[f(\boldsymbol{x}, \boldsymbol{\xi})\right], \\
U_3: &\quad \boldsymbol{x} \to \max\{\bar{f} | \mathrm{Cr}\{f(\boldsymbol{x}, \boldsymbol{\xi}) \geq \bar{f}\} \geq \beta\}, \\
U_4: &\quad \boldsymbol{x} \to H[f(\boldsymbol{x}, \boldsymbol{\xi})].
\end{aligned}
$$

Simulation for Credibility Value

According to Theorem 3.4 and the credibility inversion theorem, we know that

$$
\begin{aligned}
\mathrm{Cr}\left\{f(\boldsymbol{x}, \boldsymbol{\xi}) \geq r\right\} = &\ \frac{1}{2}\left(\sup_{x_1, x_2, \cdots, x_n \in \Re} \left\{ \min_{1 \leq i \leq n} \mu_i(x_i) \,\middle|\, f(\boldsymbol{x}, \boldsymbol{\xi}) \geq r \right\} \right. \\
&\left. + 1 - \sup_{x_1, x_2, \cdots, x_n \in \Re} \left\{ \min_{1 \leq i \leq n} \mu_i(x_i) \,\middle|\, f(\boldsymbol{x}, \boldsymbol{\xi}) < r \right\} \right).
\end{aligned}
$$

Thus we j times randomly generate real numbers u_{ij} such that $\mu_i(u_{ij}) \geq \varepsilon, i = 1, 2, \cdots, n, j = 1, 2, \cdots, N$ respectively, where ε is a sufficiently small number, and N is a sufficiently large number. Let $\boldsymbol{u}_j = (u_{1j}, u_{2j}, \cdots, u_{nj})$, and $\mu(\boldsymbol{u}_j) = \mu_{1j}(u_{1j}) \wedge \mu_{2j}(u_{2j}) \wedge \cdots \wedge \mu_{nj}(u_{nj})$. Then the credibility $\mathrm{Cr}\left\{f(\boldsymbol{x}, \boldsymbol{\xi}) \geq r\right\}$ can be obtained approximately by the following formula

$$
L = \frac{1}{2}\left(\max_{1 \leq j \leq N} \{\mu(\boldsymbol{u}_j) \mid f(\boldsymbol{x}, \boldsymbol{\xi}) \geq r\} + 1 - \max_{1 \leq j \leq N} \{\mu(\boldsymbol{u}_j) \mid f(\boldsymbol{x}, \boldsymbol{\xi}) < r\} \right),
$$

where N is a sufficiently large number.

The fuzzy simulation process for computing $\mathrm{Cr}\{f(\boldsymbol{x},\boldsymbol{\xi})\geq r\}$ is summarized as follows:

Step 1. Let $j=1$.
Step 2. Randomly generate real numbers u_{ij} such that $\mu_i(u_{ij})\geq\varepsilon, i=1,2,\cdots,n, j=1,2,\cdots,N$ respectively, where ε is a sufficiently small number, and N is a sufficiently large number.
Step 3. Set $\boldsymbol{u}_j=(u_{1j},u_{2j},\cdots,u_{nj})$, and $\mu(\boldsymbol{u}_j)=\mu_{1j}(u_{1j})\wedge\mu_{2j}(u_{2j})\wedge\cdots\wedge\mu_{nj}(u_{nj})$.
Step 4. $j\leftarrow j+1$. Turn back to Step 2 if $j\leq N$, where N is a sufficiently large number. Otherwise, turn to Step 5.
Step 5. Return L.

Simulation for Expected Value

According to the definition of expected value of fuzzy variable, we have

$$E[f(\boldsymbol{x},\boldsymbol{\xi})]=\int_0^{+\infty}\mathrm{Cr}\{f(\boldsymbol{x},\boldsymbol{\xi})\geq t\}\mathrm{d}t-\int_{-\infty}^0\mathrm{Cr}\{f(\boldsymbol{x},\boldsymbol{\xi})\leq t\}\mathrm{d}t.$$

Thus we design the procedure as follows:

Step 1. Set $E=0$.
Step 2. Randomly generate real numbers u_{ik} such that $\mu_i(u_{ij})\geq\varepsilon, i=1,2,\cdots,n, j=1,2,\cdots,N$ respectively, where ε is a sufficiently small number, and N is a sufficiently large number. Denote $\boldsymbol{u}_j=(u_{1j},u_{2j},\cdots,u_{nj})$.
Step 3. Set $a=f(\boldsymbol{x},\boldsymbol{u}_1)\wedge f(\boldsymbol{x},\boldsymbol{u}_2)\wedge\cdots\wedge f(\boldsymbol{x},\boldsymbol{u}_N)$, $b=f(\boldsymbol{x},\boldsymbol{u}_1)\vee f(\boldsymbol{x},\boldsymbol{u}_2)\vee\cdots\vee f(\boldsymbol{x},\boldsymbol{u}_N)$.
Step 4. Randomly generate t from $[a,b]$.
Step 5. If $t\geq 0$, then $E\leftarrow E+\mathrm{Cr}\{f(\boldsymbol{x},\boldsymbol{\xi})\geq t\}$.
Step 6. If $t<0$, then $E\leftarrow E-\mathrm{Cr}\{f(\boldsymbol{x},\boldsymbol{\xi})\leq t\}$.
Step 7. Repeat the fourth to sixth steps for N times, where N is a sufficiently large number.
Step 8. $E\left[f(\boldsymbol{x},\boldsymbol{\xi})\right]=a\vee 0+b\wedge 0+E\cdot(b-a)/N$.

Simulation for β-Return

In order to compute the uncertain function U_3, we randomly generate real numbers u_{ij} such that $\mu_i(u_{ij})\geq\varepsilon, i=1,2,\cdots,n, j=1,2,\cdots,N$ respectively, where ε is a sufficiently small number, and N is a sufficiently large number. Denote $\boldsymbol{u}_j=(u_{1j},u_{2j},\cdots,u_{nj})$. For any real numbers r, we set

$$\begin{aligned}D(r)=\quad&\frac{1}{2}\left(\max_{1\leq j\leq N}\{\min_{1\leq i\leq n}\boldsymbol{\mu}_i(\boldsymbol{u}_j)|f(\boldsymbol{x},\boldsymbol{u}_j)\geq r\}+\right.\\&\left.1-\max_{1\leq j\leq N}\{\min_{1\leq i\leq n}\boldsymbol{\mu}_i(\boldsymbol{u}_j)|f(\boldsymbol{x},\boldsymbol{u}_j)<r\}\right).\end{aligned}$$

Since $D(r)$ is a monotonous function of r, we may employ a bisection search to find the maximal value r such that $D(r) \geq \beta$. This value is an estimation of U_3, i.e., the β-return value. The fuzzy simulation process for computing U_3 is summarized as follows:

Step 1. Randomly generate real numbers u_{ij} such that $\mu_i(u_{ij}) \geq \varepsilon, i=1,2,\cdots,n, j = 1,2,\cdots,N$, respectively, where ε is a sufficiently small positive number, and N a sufficiently large number. Denote $\boldsymbol{u}_j = (u_{1j}, u_{2j}, \cdots, u_{nj})$.
Step 2. Find the maximal value r such that $D(r) \geq \beta$ by the bisection search.
Step 3. Return r.

Simulation for Entropy Value

According to the entropy definition, we know that

$$H[f(\boldsymbol{x}, \boldsymbol{\xi})] = \int_{-\infty}^{+\infty} S(\text{Cr}\{f(\boldsymbol{x}, \boldsymbol{\xi}) = t\})\text{d}t$$

where $S(y) = -y \ln y - (1 - y) \ln(1 - y)$.

Thus we design the fuzzy simulation procedure for calculating the entropy $H[f(\boldsymbol{x}, \boldsymbol{\xi})]$ as follows:

Step 1. Set $H = 0$.
Step 2. Randomly generate real numbers u_{ij} such that $\mu_i(u_{ij}) \geq \varepsilon, i = 1,2,\cdots,n, j = 1,2,\cdots,N$, respectively, where ε is a sufficiently small positive number, and N a sufficiently large number. Denote $\boldsymbol{u}_j = (u_{1j}, u_{2j}, \cdots, u_{nj})$.
Step 3. Set $a = \min_{1 \leq j \leq N} f(\boldsymbol{x}, \boldsymbol{u}_j)$, and $b = \max_{1 \leq j \leq N} f(\boldsymbol{x}, \boldsymbol{u}_j)$.
Step 4. Randomly generate t from $[a, b]$.
Step 5. $H \leftarrow H - (y \ln y + (1 - y) \ln(1 - y))$, where $y = \text{Cr}\{f(\boldsymbol{x}, \boldsymbol{\xi}) = t\}$.
Step 6. Repeat the fourth and fifth steps for N times, where N is a sufficiently large number.
Step 7. $H[f(\boldsymbol{x}, \boldsymbol{\xi})] = H \cdot (b - a)/N$.

Example 3.46. Let ξ_1 be a triangular fuzzy security return $(-0.1, 0.1, 0.3)$, and ξ_2 the normal fuzzy security return $\mathcal{N}(0.1, 0.1)$. Portfolio A is composed of 40% of ξ_1 and the rest 60% of ξ_2. A run of the simulation with 4000 cycles shows that the credibility value of the portfolio return not greater than 0 is 0.2490, i.e.,

$$\text{Cr}\{0.4\xi_1 + 0.6\xi_2 \leq 0\} = 0.2490.$$

The simulation procedures are as follows:

Step 1. Let $j = 1$.
Step 2. Randomly generate real numbers a from $(-0.1, 0.3)$ and b from $(-0.4, 0.6)$ (we generate b from $(-0.4, 0.6)$ because $\mu(t) \approx 0$ when $t < e - 5\sigma$, and $\mu(t) \approx 0$ when $t > e + 5\sigma$, where e is the expected value and σ the positive square root of variance of the normal fuzzy variable).

Step 3. Calculate $\mu(a) = \dfrac{a+0.1}{0.1+0.1}$ if $a \le 0.1$, and $\mu(a) = \dfrac{0.3-a}{0.3-0.1}$ if $a > 0.1$, and $\mu(b) = 2\left(1+\exp\left(\dfrac{\pi|b-0.1|}{\sqrt{6}\times 0.1}\right)\right)^{-1}$.
Step 4. Set $\nu_j = \mu(a) \wedge \mu(b)$.
Step 5. $j \leftarrow j+1$. Turn back to Step 2 if $j \le 4000$. Otherwise, turn to Step 6.
Step 6. Return

$$L = \frac{1}{2}\left(\max_{1\le j\le N}\left\{\nu_j \mid 0.4a+0.6b \le 0\right\} + 1 - \max_{1\le j\le N}\left\{\nu_j \mid 0.4a+0.6b > 0\right\}\right).$$

Example 3.47. For the above mentioned portfolio A, assume the investors set the confidence level $\beta = 0.9$. A run of the simulation with 3000 cycles shows that the 0.9-return value of the portfolio is -0.0591, i.e.,

$$\xi(0.9) = \sup\{\bar{f}|\text{Cr}\{0.4\xi_1 + 0.6\xi_2 \ge \bar{f}\} \ge 0.9\} = -0.0591.$$

The simulation procedures are as follows:

Step 1. Randomly generate real numbers a_j from $(-0.1, 0.3)$ and b_j from $(-0.4, 0.6)$ for $j = 1, 2, \cdots, 3000$.
Step 2. Calculate $\mu(a_j) = \dfrac{a+0.1}{0.1+0.1}$ if $a_j \le 0.1$, and $\mu(a_j) = \dfrac{0.3-a}{0.3-0.1}$ if $a_j > 0.1$, and $\mu(b_j) = 2\left(1+\exp\left(\dfrac{\pi|b_j-0.1|}{\sqrt{6}\times 0.1}\right)\right)^{-1}$.
Step 3. Set $\nu_j = \mu(a_j) \wedge \mu(b_j)$.
Step 4. Let

$$D(r) = \frac{1}{2}\left(\max_{1\le j\le N}\left\{\nu_j \mid 0.4a+0.6b \le r\right\} + 1 - \max_{1\le j\le N}\left\{\nu_j \mid 0.4a+0.6b > r\right\}\right).$$

Find the maximal value r such that $D(r) \ge 0.9$ by the bisection search.
Step 5. Return r.

Example 3.48. Let ξ_1 be a triangular fuzzy security return $(-0.2, 0, 0.4)$, and ξ_2 the normal fuzzy security return $\mathcal{N}(0.1, 0.1)$. Portfolio A is composed of 40% of ξ_1 and the rest 60% of ξ_2. A run of the simulation with 8000 cycles shows that the semivariance value of the portfolio, i.e.,

$$SV[0.4\xi_1 + 0.6\xi_2] = 0.0432.$$

The simulation procedures are as follows:

Step 1. Randomly 8000 times generate real numbers a_i from $(-0.2, 0.4)$ and b_i from $(-0.4, 0.6), i = 1, 2, \cdots, 8000$.
Step 2. If $0.4a_i + 0.6b_i - 0.08 \le 0$ (the expected value of Portfolio A is 0.08), set $r_i = (0.4a_i + 0.6b_i - 0.08)^2$; otherwise, set $r_i = 0$.

Step 3. Set $a = r_1 \wedge r_2 \wedge \cdots \wedge r_{8000}$ and $b = r_1 \vee r_2 \vee \cdots \vee r_{8000}$.
Step 4. Set $V = 0$.
Step 5. Randomly generate t from $[a, b]$.
Step 6. If $t \geq 0$, then $V \leftarrow V + \mathrm{Cr}\{(0.4\xi_1 + 0.6\xi_2 - 0.08)^2 \geq t\}$; if $t < 0$, then $V \leftarrow V + 0$.
Step 7. Repeat the fifth to sixth steps 8000 times.
Step 6. $V[0.4\xi_1 + 0.6\xi_2] = a \vee 0 + b \wedge 0 + V \cdot (b - a)/8000$.

Example 3.49. For the above mentioned portfolio A, a run of the simulation with 8000 cycles shows that the variance value of the portfolio, i.e.,

$$V[0.4\xi_1 + 0.6\xi_2] = -0.0710.$$

3.7.2 Hybrid Intelligent Algorithm

When the objective and constraint values have been calculated by fuzzy simulation, simulation results are integrated into the GA introduced in Subsection 2.6.3 to produce a hybrid intelligent algorithm. After selection, crossover and mutation, the new population is ready for its next evaluation. The hybrid intelligent algorithm will continue until a given number of cyclic repetitions of the above steps is met. We summarize the algorithm as follows:

Step 1. Initialize *pop_size* chromosomes.
Step 2. Calculate the objective values for all chromosomes by fuzzy simulation.
Step 3. Give the rank order of the chromosomes according to the objective values, and compute the values of the rank-based evaluation function of the chromosomes.
Step 4. Compute the fitness of each chromosome according to the rank-based-evaluation function.
Step 5. Select the chromosomes by spinning the roulette wheel.
Step 6. Update the chromosomes by crossover and mutation operations.
Step 7. Repeat the second to the sixth steps for a given number of cycles.
Step 8. Take the best chromosome as the solution of portfolio selection.

3.7.3 Numerical Example

Suppose an investor adopts credibility minimization selection idea and wants to choose an optimal portfolio from ten securities of which five security return rates are normal fuzzy variables and the rest five the triangular fuzzy variables. The prediction of the return rates of the ten securities is given in Table 3.10. Suppose the minimum expected return the investor can accept is 0.031, and the investor wants to minimize the occurrence credibility of portfolio return below a disaster level -0.08. According to the credibility minimization selection idea introduced in Subsection 3.4.1, we build the model as follows:

$$\begin{cases} \min \mathrm{Cr}\{\xi_1 x_1 + \xi_2 x_2 + \cdots + \xi_{10} x_{10} \le -0.08\} \\ \text{subject to:} \\ \qquad E[\xi_1 x_1 + \xi_2 x_2 + \cdots + \xi_{10} x_{10}] \ge 0.031 \\ \qquad x_1 + x_2 + \cdots + x_{10} = 1 \\ \qquad x_i \ge 0, \quad i = 1, 2, \cdots, 10. \end{cases} \tag{3.70}$$

The constraint $E[\xi_1 x_1 + \xi_2 x_2 + \cdots + \xi_{10} x_{10}]$ can be calculated via

$$\sum_{i=1}^{5} e_i x_i + \sum_{i=6}^{10} \frac{1}{4}(a_i x_i + 2b_i x_i + c_i x_i).$$

The objective (i.e., credibility value) is calculated by fuzzy simulation. Then the simulation result is integrated into the GA introduced in Subsection 2.6.3 to produce the hybrid intelligent algorithm. A run of the algorithm with 10000 generations shows that in order to minimize the credibility of portfolio return not greater than -0.08 with the constraint that the expected return of the portfolio should not be less than 0.031, the investor should allocate his/her money according to Table 3.11. The minimum credibility level of portfolio

Table 3.10 Fuzzy Return Rates of Ten Securities

Security i	$\xi_i \sim \mathcal{N}(e_i, \sigma_i)$	Security i	$\xi_i = (a_i, b_i, c_i)$
1	$\mathcal{N}(0.033, 0.12)$	6	$(-0.08, 0.026, 0.09)$
2	$\mathcal{N}(0.032, 0.11)$	7	$(-0.09, 0.030, 0.10)$
3	$\mathcal{N}(0.033, 0.14)$	8	$(-0.15, 0.032, 0.16)$
4	$\mathcal{N}(0.031, 0.11)$	9	$(-0.12, 0.04, 0.10)$
5	$\mathcal{N}(0.025, 0.07)$	10	$(-0.12, 0.05, 0.12)$

Table 3.11 Allocation of Money to Ten Securities

Security i	1	2	3	4	5
Allocation of money	0.00%	85.90%	0.00%	0.00%	0.00%
Security i	6	7	8	9	10
Allocation of money	0.00%	0.00 %	0.00%	0.00%	14.10%

return not greater than -0.08 is 0.1805. The hybrid intelligent algorithm is summarized below.

Hybrid Intelligent Algorithm:

Step 1. Determine representation structure of solutions by chromosomes. In the example, the genes $c_1, c_2, \cdots, c_{10}$ in a chromosome $C = (c_1, c_2, \cdots, c_n)$ are restricted in the interval $[0, 1]$. A solution $\boldsymbol{x} = (x_1, x_2, \cdots, x_{10})$ is matched with a chromosome in the following way,

$$x_i = \frac{c_i}{c_1 + c_2 + \cdots + c_{10}}, \quad i = 1, 2, \cdots, n$$

which ensures that $x_1 + x_2 + \cdots + x_{10} = 1$ always holds.

Step 2. Set parameters $P_c = 0.3, P_m = 0.2, pop\_size = 30$ in the GA.

Step 3. Generate the chromosomes $C = (c_1, c_2, \cdots, c_{10})$ from $[0, 1]^{10}$.

Step 4. Calculate the expected return for each chromosome according to the formula

$$\sum_{i=1}^{5} e_i x_i + \sum_{i=6}^{10} \frac{1}{4}(a_i x_i + 2b_i x_i + c_i x_i).$$

Then check the feasibility of the chromosome as follows:

If $\sum_{i=1}^{5} e_i x_i + \sum_{i=6}^{10} \frac{1}{4}(a_i x_i + 2b_i x_i + c_i x_i) \geq 0.031$

return 1;

return 0;

in which 1 means feasible, and 0 non-feasible.

Step 5. Repeat the third and fourth steps until feasible *pop_size* numbers of chromosomes are produced.

Step 6. Calculate the objective values (i.e., credibility values) via fuzzy simulation and give the rank order of the chromosomes according to the objective values to make the better chromosomes take the smaller ordinal numbers.

Step 7. Compute the values of the rank-based evaluation function for all the chromosomes.

Step 8. Calculate the fitness of each chromosome according to the rank-based-evaluation function.

Step 9. Select the chromosomes by spinning the roulette wheel.

Step 10. Update the chromosomes by crossover and mutation operations.

Step 11. Repeat the sixth to tenth steps for 10000 cycles.

Step 12. Take the best chromosome as the solution of the portfolio selection problem.

Chapter 4
Uncertain Portfolio Selection

Though randomness and fuzziness are two basic types of uncertain phenomena, uncertainty in real life is varied. Sometimes, uncertainty behaves neither randomly nor fuzzily. For example, the occurrence chance of a security price falling in the interval of [100, 110] is 30%, and the occurrence chance of the security price in the interval of [110, 120] is 20%. Then what is the occurrence chance of the security price in the interval of [100, 120]? A survey shows that some people believe that the occurrence chance should be in somewhere that is not less than 30% but not greater than 50%. In this case, the security price is neither random nor fuzzy. Uncertainty theory which is founded by Liu [60] in 2007 provides a new tool to handle this type of uncertainty. Uncertain portfolio selection, a new topic introduced by Huang [42], deals with portfolio selection by means of uncertainty theory when portfolio return is neither random nor fuzzy.

This chapter will first introduce some fundamentals about uncertainty theory concerning uncertain portfolio selection. Then based on different definitions of risk, we will introduce a spectrum of uncertain portfolio selection models. Methods for solving the models will also be provided.

4.1 Fundamentals of Uncertainty Theory

Uncertainty theory is a branch of mathematics based on normality, monotonicity, self-duality, countable subadditivity, and product measure axioms for studying the uncertain phenomenon which is neither random nor fuzzy. *Uncertain measure* which is the core of the uncertainty theory is used to measure the truth value of an uncertain event.

Uncertain Measure and Uncertainty Space

Definition 4.1. *(Liu [60]) Let Γ be a nonempty set, and $\mathcal{L}$ a σ-algebra over Γ. Each element $\Lambda \in \mathcal{L}$ is called an event. To each event Λ a number $\mathcal{M}\{\Lambda\}$ indicates the level that Λ will occur. The set function $\mathcal{M}$ is called an uncertain measure if it satisfies the following four axioms*

X. Huang: Portfolio Analysis: From Probab. to Credibilistic, STUDFUZZ 250, pp. 117–156.
springerlink.com

(Axiom 1) (Normality) $\mathcal{M}\{\Gamma\} = 1$.
(Axiom 2) (Monotonicity) $\mathcal{M}\{\Lambda_1\} \leq \mathcal{M}\{\Lambda_1\}$ *whenever* $\Lambda_1 \subset \Lambda_2$.
(Axiom 3) (Self-Duality) $\mathcal{M}\{\Lambda\} + \mathcal{M}\{\Lambda^c\} = 1$.
(Axiom 4) (Countable Subadditivity) For every countable sequence of events $\{\Lambda_i\}$, *we have*

$$\mathcal{M}\left\{\bigcup_{i=1}^{\infty} \Lambda_i\right\} \leq \sum_{i=1}^{\infty} \mathcal{M}\{\Lambda_i\}.$$

The triplet $(\Gamma, \mathcal{L}, \mathcal{M})$ *is called an uncertainty space.*

Remark 4.1. Self-duality is consistent with the law of contradiction and the law of excluded middle. In Question 2 in page 66 in Chapter 3, we have shown that without self-duality, a measure would be rather strange because the judgement made based on the measure without self-duality property would be in contradiction with the law of contradiction and the law of excluded middle.

Remark 4.2. A measure would be rather strange if it has no subadditivity property. For example [64], assume we have a universal set consisting of three elements. Let us define a set function that takes value 0 for each singleton and value 1 for each set with at least two elements. Then such a set function satisfies axioms 1, 2 and 3 but does not satisfy subadditivity. It is seen that pathology would occur if we used such a kind of set function as a measure.

Remark 4.3. A measure would be rather strange if countable subadditivity axiom were replaced by finite subadditivity axiom. For example [64], assume we have a universal set consisting of all real numbers. Let us define a set function that takes value 0 if the set is bounded, 0.5 if both the set and complement are unbounded, and 1 if the complement of the set is bounded. Then such a set function satisfies axioms 1, 2, 3 and finite subadditivity instead of countable subadditivity. It is seen that pathology would occur if we used such a kind of set function as a measure.

Remark 4.4. Let Γ be a nonempty set, $\mathcal{L}$ a σ-algebra over Γ, and $\mathcal{M}$ an uncertain measure. From Axioms 1 and 3 we know $\mathcal{M}\{\emptyset\} = 0$. From axiom 2 we know $0 \leq \mathcal{M}\{\Lambda\} \leq 1$ for any $\Lambda \in \mathcal{L}$ because $\emptyset \subset \Lambda \subset \Gamma$. That is, the value of an uncertain measure of an uncertain event is in the interval $[0, 1]$.

Product Measure Axiom

Product uncertain measure and Product Measure Axiom was defined by Liu [63] in 2009.

Axiom 5. *(Product Measure Axiom, Liu [63]) Let* $(\Gamma_k, \mathcal{L}_k, \mathcal{M}_k)$ *be uncertainty spaces for* $k = 1, 2, \cdots, n$. *Write*

$$\Gamma = \Gamma_1 \times \Gamma_2 \times \cdots \times \Gamma_n, \quad \mathcal{L} = \mathcal{L}_1 \times \mathcal{L}_2 \times \cdots \times \mathcal{L}_n.$$

Then the product uncertain measure on Γ is

$$\mathcal{M}\{\Lambda\} = \begin{cases} \sup\limits_{\Lambda_1 \times \Lambda_2 \times \cdots \times \Lambda_n \subset \Lambda} \min\limits_{1 \le k \le n} \mathcal{M}_k\{\Lambda_k\}, \\ \quad \text{if} \sup\limits_{\Lambda_1 \times \Lambda_2 \times \cdots \times \Lambda_n \subset \Lambda} \min\limits_{1 \le k \le n} \mathcal{M}_k\{\Lambda_k\} > 0.5 \\ 1 - \sup\limits_{\Lambda_1 \times \Lambda_2 \times \cdots \times \Lambda_n \subset \Lambda^c} \min\limits_{1 \le k \le n} \mathcal{M}_k\{\Lambda_k\}, \\ \quad \text{if} \sup\limits_{\Lambda_1 \times \Lambda_2 \times \cdots \times \Lambda_n \subset \Lambda^c} \min\limits_{1 \le k \le n} \mathcal{M}_k\{\Lambda_k\} > 0.5 \\ 0.5, \quad \text{otherwise} \end{cases} \tag{4.1}$$

for each $\Lambda \in \mathcal{L}$.

Theorem 4.1. *(Peng [75]) The product uncertain measure defined by Equation (4.1) is an uncertain measure.*

Uncertain Variable

Definition 4.2. *(Liu [60]) An uncertain variable is defined as a measurable function from an uncertainty space $(\Gamma, \mathcal{L}, \mathcal{M})$ to the set of real numbers.*

Remark 4.5. Since an uncertain variable ξ is a measurable function, for any Borel set B of real numbers, the set

$$\{\xi \in B\} = \{\gamma \in \Gamma | \xi(\gamma) \in B\} \in \mathcal{L},$$

which means that $\{\xi \in B\}$ is an event. In practice, we usually express the event $\{\xi \in B\}$ by $\{\xi \le t\}$ or $\{\xi \ge t\}$ where t is a real number. For example, let ξ represent an uncertain portfolio return. Then the event that the portfolio return is not less than 0.10 can be expressed by $\{\xi \ge 0.10\}$.

As a practitioner, we are not interested in the specific nature of the sample space Γ nor the specific function which defines the uncertain variable ξ. Instead, we are interested in the values of the uncertain measure of the uncertain variable taking some real values, for example, $\mathcal{M}\{\gamma \in \Gamma | \xi(\gamma) \le t\}$, or simply $\mathcal{M}\{\xi \le t\}$. In uncertain portfolio selection, we are also only interested in the values of the uncertain measure that the portfolio return takes certain values.

Definition 4.3. *Let ξ_1 and ξ_2 be uncertain variables defined on the uncertainty space $(\Gamma, \mathcal{L}, \mathcal{M})$. We say $\xi_1 = \xi_2$ if $\xi_1(\gamma) = \xi_2(\gamma)$ for almost all $\gamma \in \Gamma$.*

Uncertainty Distribution

Definition 4.4. *(Liu [60]) The uncertainty distribution $\Phi : \Re \to [0,1]$ of an uncertain variable ξ is defined by*

$$\Phi(t) = \mathcal{M}\{\xi \le t\}. \tag{4.2}$$

Remark 4.6. An uncertain variable has a unique uncertainty distribution function, but an uncertainty distribution function may produce multiple uncertain variables. For example, let $\Gamma = \{\gamma_1, \gamma_2\}$ and $\mathcal{M}\{\gamma_1\} = \mathcal{M}\{\gamma_2\} = 0.5$. Then $(\Gamma, \mathcal{L}, \mathcal{M})$ is an uncertainty space. Define two uncertain variables

$$\xi_1(\gamma) = \begin{cases} -1, & \text{if} \quad \gamma = \gamma_1 \\ 1, & \text{if} \quad \gamma = \gamma_2, \end{cases} \qquad \xi_2(\gamma) = \begin{cases} 1, & \text{if} \quad \gamma = \gamma_1 \\ -1, & \text{if} \quad \gamma = \gamma_2. \end{cases}$$

We can find that ξ_1 and ξ_2 have the same uncertainty distribution, i.e.,

$$\Phi(t) = \begin{cases} 0, & \text{if} \quad t < -1 \\ 0.5, & \text{if} \quad -1 \le t < 1 \\ 1, & \text{if} \quad t \ge 1. \end{cases}$$

However, it is clear that ξ_1 and ξ_2 are not the same uncertain variable in the sense of Definition 4.3. Since one uncertainty distribution function may produce multiple uncertain variables, we can not define an uncertain variable via distribution function. An axiomatic system is needed to define an uncertain variable and to discuss the properties concerning the uncertain variable such that the discussion is precise and consistent. However, in application, it is enough for us to start with uncertainty distributions to study the behavior of the uncertain variables and make our decisions.

Example 4.1. Let ξ be an uncertain variable with uncertainty distribution Φ. Then for any number $k > 0$, the uncertainty distribution of $k\xi$

$$\Psi(t) = \Phi\left(\frac{t}{k}\right) \quad \text{and} \quad \Psi^{-1}(\alpha) = k\Phi^{-1}(\alpha). \tag{4.3}$$

Theorem 4.2. *(Peng and Iwamura [77], Sufficient and Necessary Condition for Uncertainty Distribution) A function $\Phi : \Re \to [0,1]$ is an uncertainty distribution if and only if it is an increasing function except $\Phi(t) \equiv 0$ and $\Phi(t) \equiv 1$.*

Two Special Uncertain Variables

Definition 4.5. *An uncertain variable ξ is called a linear uncertain variable if it has a linear uncertainty distribution function*

$$\Phi(t) = \begin{cases} 0, & \text{if } t \le a \\ \dfrac{t-a}{b-a}, & \text{if } a \le t \le b \\ 1, & \text{otherwise.} \end{cases}$$

We denote it by $\xi = \mathcal{L}(a, b)$ where a and b are real numbers and $a < b$.

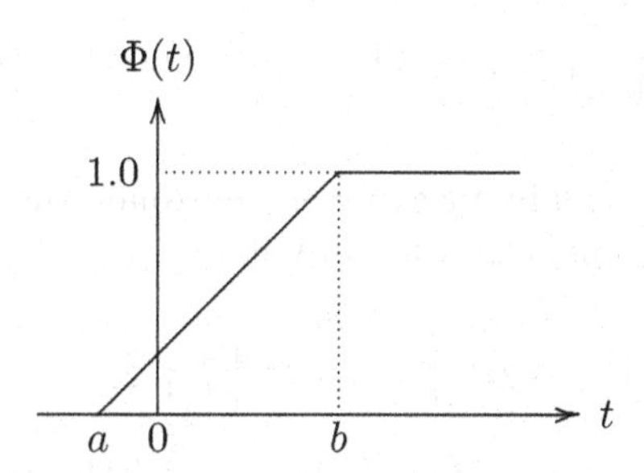

Fig. 4.1 Uncertainty distribution of a linear uncertain variable.

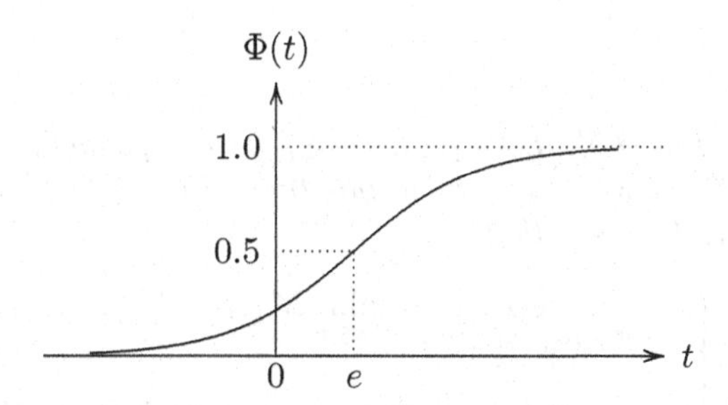

Fig. 4.2 Uncertainty distribution of a normal uncertain variable.

Definition 4.6. *An uncertain variable ξ is called a normal uncertain variable if it has a normal uncertainty distribution function*

$$\Phi(t) = \left(1 + \exp\left(\frac{\pi(e-t)}{\sqrt{3}\sigma}\right)\right)^{-1}, \quad t \in R, \quad \sigma > 0.$$

We denoted it by $\xi \sim \mathcal{N}(e, \sigma)$.

Independence

Definition 4.7. *(Liu [63]) The uncertain variables $\xi_1, \xi_2, \cdots, \xi_n$ are said to be independent if*

$$\mathcal{M}\left\{\bigcap_{i=1}^{n}\{\xi_i \in B_i\}\right\} = \min_{1\le i\le n} \mathcal{M}\{\xi_i \in B_i\} \tag{4.4}$$

for any Borel sets $B_1, B_2, \cdots, B_n$ of real numbers.

Theorem 4.3. *(Liu [64]) The uncertain variables $\xi_1, \xi_2, \cdots, \xi_n$ are independent if and only if*

$$\mathcal{M}\left\{\bigcup_{i=1}^{n}\{\xi_i \in B_i\}\right\} = \max_{1\le i\le n} \mathcal{M}\{\xi_i \in B_i\}. \tag{4.5}$$

Proof: Since the uncertain measure is self-dual, the uncertain variables $\xi_1, \xi_2, \cdots, \xi_n$ are independent if and only if

$$\begin{aligned}\mathcal{M}\left\{\bigcup_{i=1}^{n}\{\xi_i \in B_i\}\right\} &= 1 - \mathcal{M}\left\{\bigcap_{i=1}^{n}\{\xi_i \in B_i^c\}\right\} \\ &= 1 - \min_{1\le i\le n} \mathcal{M}\{\xi_i \in B_i^c\} = \max_{1\le i\le n} \mathcal{M}\{\xi_i \in B_i\}.\end{aligned}$$

Thus, the theorem is proven.

Operational Law

Theorem 4.4. *(Liu [63]) Let $\xi_1, \xi_2, \cdots, \xi_n$ be independent uncertain variables, and $f : \Re^n \to \Re$ a measurable function. Then $\xi = f(\xi_1, \xi_2, \cdots, \xi_n)$ is an uncertain variable such that*

$$\mathcal{M}\{\xi \in B\} = \begin{cases} \sup\limits_{f(B_1,B_2,\cdots,B_n)\subset B} \min\limits_{1\le k\le n} \mathcal{M}\{\xi_k \in B_k\}, \\ \qquad \text{if } \sup\limits_{f(B_1,B_2,\cdots,B_n)\subset B} \min\limits_{1\le k\le n} \mathcal{M}\{\xi_k \in B_k\} > 0.5 \\ 1 - \sup\limits_{f(B_1,B_2,\cdots,B_n)\subset B^c} \min\limits_{1\le k\le n} \mathcal{M}\{\xi_k \in B_k\}, \\ \qquad \text{if } \sup\limits_{f(B_1,B_2,\cdots,B_n)\subset B^c} \min\limits_{1\le k\le n} \mathcal{M}\{\xi_k \in B_k\} > 0.5 \\ 0.5, \qquad \text{otherwise} \end{cases} \tag{4.6}$$

where $B_1, B_2, \cdots, B_n, B$ are Borel sets of real numbers, and $f(B_1, B_2, \cdots, B_n) \subset B$ means $f(t_1, t_2, \cdots, t_n) \subset B$ for any $t_1 \in B_1, t_2 \in B_2, \cdots, t_n \in B_n$.

Proof: The theorem can be proven directly from the product measure axiom.

Theorem 4.5. *(Liu [64]) Let ξ be an uncertain variable with uncertainty distribution Φ, and let f be a strictly increasing function. Then the uncertainty distribution of $f(\xi)$ can be obtained via*

$$\Psi(t) = \Phi(f^{-1}(t)) \tag{4.7}$$

which can also be expressed by

$$\Psi^{-1}(\alpha) = f\left(\Phi^{-1}(\alpha)\right), \quad 0 < \alpha < 1. \tag{4.8}$$

Proof: For any real numbers t, since f is a strictly increasing function, we have

$$f((-\infty, f^{-1}(t)]) = (-\infty, t].$$

According to the operational law, we have

$$\Psi(t) = \mathcal{M}\{f(\xi) \in (-\infty, t]\} = \mathcal{M}\{\xi \in (-\infty, f^{-1}(t)]\} = \Phi(f^{-1}(t)).$$

Theorem 4.6. *(Peng [76]) Let $\xi_1, \xi_2, \cdots, \xi_n$ be independent uncertain variables with continuous uncertainty distributions $\Phi_1, \Phi_2, \cdots, \Phi_n$, respectively, and Ψ the uncertainty distribution of the sum $\xi_1 + \xi_2 + \cdots + \xi_n$. If $\Phi_1^{-1}(\alpha)$, $\Phi_2^{-1}(\alpha), \cdots, \Phi_n^{-1}(\alpha)$ are unique for each $\alpha \in (0, 1)$, we have*

$$\Psi^{-1}(\alpha) = \Phi_1^{-1}(\alpha) + \Phi_2^{-1}(\alpha) + \cdots + \Phi_n^{-1}(\alpha), \quad 0 < \alpha < 1. \tag{4.9}$$

Proof: According to monotonicity property of uncertain measure, for any given $\alpha \in (0, 1)$, we have

$$\mathcal{M}\left\{\sum_{i=1}^{n} \xi_i \le \sum_{i=1}^{n} \Phi_i^{-1}(\alpha)\right\} \ge \mathcal{M}\left\{\bigcap_{i=1}^{n} \left(\xi_i \le \Phi_i^{-1}(\alpha)\right)\right\}.$$

Since $\xi_1, \xi_2, \cdots, \xi_n$ are independent uncertain variables, according to Equation (4.4), we have

$$\mathcal{M}\left\{\sum_{i=1}^{n} \xi_i \le \sum_{i=1}^{n} \Phi_i^{-1}(\alpha)\right\} \ge \mathcal{M}\left\{\bigcap_{i=1}^{n} \left(\xi_i \le \Phi_i^{-1}(\alpha)\right)\right\}$$
$$= \min_{1 \le i \le n} \mathcal{M}\{\xi_i \le \Phi_i^{-1}(\alpha)\} = \min_{1 \le i \le n} \alpha = \alpha.$$

On the other hand, for any number $\epsilon > 0$, we have

$$\mathcal{M}\left\{\sum_{i=1}^{n} \xi_i \le \sum_{i=1}^{n} \Phi_i^{-1}(\alpha) - \epsilon\right\} \le \mathcal{M}\left\{\bigcup_{i=1}^{n} \left(\xi_i \le \Phi_i^{-1}(\alpha) - \frac{\epsilon}{n}\right)\right\}$$

because uncertain measure is monotonous. Since $\xi_1, \xi_2, \cdots, \xi_n$ are independent uncertain variables, according to Equation (4.5), we have

$$\mathcal{M}\left\{\sum_{i=1}^{n} \xi_i \le \sum_{i=1}^{n} \Phi_i^{-1}(\alpha) - \epsilon\right\} \le \mathcal{M}\left\{\bigcup_{i=1}^{n} \left(\xi_i \le \Phi_i^{-1}(\alpha) - \frac{\epsilon}{n}\right)\right\}$$
$$= \max_{1 \le i \le n} \mathcal{M}\left\{\xi_i \le \Phi_i^{-1}(\alpha) - \frac{\epsilon}{n}\right\} < \max_{1 \le i \le n} \alpha = \alpha.$$

It follows from the continuity of uncertainty distributions that

$$\mathcal{M}\{\xi_1 + \xi_2 + \cdots + \xi_n \le \Phi_1^{-1}(\alpha) + \Phi_2^{-1}(\alpha) + \cdots + \Phi_n^{-1}(\alpha)\} = \alpha$$

which implies that

$$\Psi^{-1}(\alpha) = \Phi_1^{-1}(\alpha) + \Phi_2^{-1}(\alpha) + \cdots + \Phi_n^{-1}(\alpha).$$

Equation (4.9) is proven.

9999 Method A. *Let ξ_i be uncertain variables with uncertainty distributions Φ_i, and k_i any positive numbers, $i = 1, 2, \cdots, n$, respectively. Let Ψ_i represent the uncertainty distributions of $k_i\xi_i, i = 1, 2, \cdots, n$, respectively. Then according to Theorems 4.5, we have*

$$\Psi_i^{-1}(\alpha) = k_i\Phi_i^{-1}(\alpha).$$

Let Ψ represent the uncertainty distribution of $k_1\xi_1 + k_2\xi_2 + \cdots + k_n\xi_n$. Then according to Theorem 4.6, we have

$$\Psi^{-1}(\alpha) = \sum_{i=1}^{n} \Psi_i^{-1}(\alpha) = \sum_{i=1}^{n} k_i\Phi_i^{-1}(\alpha).$$

That is, the uncertainty distribution Ψ of $k_1\xi_1 + k_2\xi_2 + \cdots + k_n\xi_n$ can be represented on a computer as follows:

α_i	0.0001	0.0002	0.0003	$\cdots$	0.9999
$\Phi_1^{-1}(\alpha_i)$	$t_{1/1}$	$t_{1/2}$	$t_{1/3}$	$\cdots$	$t_{1/9999}$
$\Phi_2^{-1}(\alpha_i)$	$t_{2/1}$	$t_{2/2}$	$t_{2/3}$	$\cdots$	$t_{2/9999}$
$\Phi_3^{-1}(\alpha_i)$	$t_{3/1}$	$t_{3/2}$	$t_{3/3}$	$\cdots$	$t_{3/9999}$
$\cdots$	$\cdots$	$\cdots$	$\cdots$	$\cdots$	$\cdots$
$\Phi_n^{-1}(\alpha_i)$	$t_{n/1}$	$t_{n/2}$	$t_{n/3}$	$\cdots$	$t_{n/9999}$
$\Psi^{-1}(\alpha_i)$	$\sum_{i=1}^{n} k_i t_{i/1}$	$\sum_{i=1}^{n} k_i t_{i/2}$	$\sum_{i=1}^{n} k_i t_{i/3}$	$\cdots$	$\sum_{i=1}^{n} k_i t_{i/9999}$

(4.10)

Remark 4.7. According to the precision requirement of the researcher, 9999 Method can also be a 99 Method or 999999 Method. For example, if the precision is required to be higher, the above introduced 9999 Method becomes 999999 Method, and the uncertainty distribution of $k_1\xi_1 + k_2\xi_2 + \cdots + k_n\xi_n$ can be obtained as follows:

α_i	0.000001	0.000002	0.000003	$\cdots$	0.999999
$\Psi^{-1}(\alpha_i)$	$\sum_{i=1}^{n} k_i t_{i/1}$	$\sum_{i=1}^{n} k_i t_{i/2}$	$\sum_{i=1}^{n} k_i t_{i/3}$	$\cdots$	$\sum_{i=1}^{n} k_i t_{i/999999}$

Theorem 4.7. *Let ξ_1 and ξ_2 be two independent linear uncertain variables $\mathcal{L}(a_1,b_1)$ and $\mathcal{L}(a_2,b_2)$, respectively. Then the sum $\xi_1+\xi_2$ is also a linear uncertain variable, and*

$$\mathcal{L}(a_1,b_1)+\mathcal{L}(a_2,b_2)=\mathcal{L}(a_1+a_2,b_1+b_2). \tag{4.11}$$

The product of a linear uncertain variable $\mathcal{L}(a,b)$ and a scalar number $k>0$ is also a linear uncertain variable, and

$$k\cdot\mathcal{L}(a,b)=\mathcal{L}(ka,kb). \tag{4.12}$$

Proof: Suppose that Φ_1 and Φ_2 are uncertainty distributions of linear uncertain variables ξ_1 and ξ_2, respectively. Then we have

$$\Phi_1^{-1}(\alpha)=(1-\alpha)a_1+\alpha b_1,$$
$$\Phi_2^{-1}(\alpha)=(1-\alpha)a_2+\alpha b_2.$$

According to Theorem 4.6, the uncertainty distribution Ψ of $\xi_1+\xi_2$ can be expressed by

$$\Psi^{-1}(\alpha)=\Phi_1^{-1}(\alpha)+\Phi_2^{-1}(\alpha)=(1-\alpha)(a_1+a_2)+\alpha(b_1+b_2),$$

which implies that the sum of linear uncertain variable is also a linear uncertain variable and

$$\mathcal{L}(a_1,b_1)+\mathcal{L}(a_2,b_2)=\mathcal{L}(a_1+a_2,b_1+b_2).$$

Equation (4.11) is proven.

For a linear uncertain variable $\mathcal{L}(a,b)$, according to Theorem 4.5, when $k>0$, the uncertainty distribution Ψ of $k\xi$ is

$$\Psi(t)=\Phi\left(\frac{t}{k}\right)=\frac{t/k-a}{b-a}=\frac{t-ka}{k(b-a)}.$$

Thus, Equation (4.12) is proven.

Theorem 4.8. *Let ξ_1 and ξ_2 be two independent normal uncertain variables $\mathcal{N}(e_1,\sigma_1)$ and $\mathcal{N}(e_2,\sigma_2)$, respectively. Then the sum $\xi_1+\xi_2$ is also a normal uncertain variable, and*

$$\mathcal{N}(e_1,\sigma_1)+\mathcal{N}(e_2,\sigma_2)=\mathcal{N}(e_1+e_2,\sigma_1+\sigma_2). \tag{4.13}$$

The product of a normal uncertain variable $\mathcal{N}(e,\sigma)$ and a scalar number k is also a normal uncertain variable, and

$$k\cdot\mathcal{N}(e,\sigma)=\mathcal{N}(ke,|k|\sigma). \tag{4.14}$$

Proof: Suppose that Φ_1 and Φ_2 are uncertainty distributions of normal uncertain variables ξ_1 and ξ_2, respectively. Then we have

$$\Phi_1^{-1}(\alpha) = e_1 + \frac{\sqrt{3}\sigma_1}{\pi}\ln\frac{\alpha}{1-\alpha},$$
$$\Phi_2^{-1}(\alpha) = e_2 + \frac{\sqrt{3}\sigma_2}{\pi}\ln\frac{\alpha}{1-\alpha}.$$

According to Theorem 4.6, the uncertainty distribution Ψ of $\xi_1 + \xi_2$ can be expressed by

$$\Psi^{-1}(\alpha) = \Phi_1^{-1}(\alpha) + \Phi_2^{-1}(\alpha) = (e_1 + e_2) + \frac{\sqrt{3}(\sigma_1 + \sigma_2)}{\pi}\ln\frac{\alpha}{1-\alpha},$$

which implies that the sum of normal uncertain variable is also a normal uncertain variable, and Equation (4.13) holds.

For a normal uncertain variable $\mathcal{N}(e,\sigma)$, according to Theorem 4.6, when $k > 0$, the uncertainty distribution Ψ of $k\xi$ is

$$\Psi(t) = \Phi\left(\frac{t}{k}\right) = \left(1 + \exp\left(\frac{\pi(e - t/k)}{\sqrt{3}\sigma}\right)\right)^{-1} = \left(1 + \exp\left(\frac{\pi(ke - t)}{\sqrt{3}k\sigma}\right)\right)^{-1}.$$

Thus, Equation (4.14) holds.

Expected Value

Definition 4.8. *(Liu [60]) Let ξ be an uncertain variable. Then the expected value of ξ is defined by*

$$E[\xi] = \int_0^{+\infty} \mathcal{M}\{\xi \ge t\}\mathrm{d}t - \int_{-\infty}^0 \mathcal{M}\{\xi \le t\}\mathrm{d}t \tag{4.15}$$

provided that at least one of the two integrals is finite.

Example 4.2. Let ξ be a linear uncertain variable $\mathcal{L}(a,b)$. Then its uncertainty distribution is

$$\Phi(t) = \begin{cases} 0, & \text{if } t \le a \\ \dfrac{t-a}{b-a}, & \text{if } a \le t \le b \\ 1, & \text{otherwise.} \end{cases}$$

Thus, if $a \ge 0$, its expected value is

$$E[\xi] = \left(\int_0^a 1\mathrm{d}t + \int_a^b \left(1 - \frac{t-a}{b-a}\right)\mathrm{d}t + \int_b^{+\infty} 0\mathrm{d}t\right) - \int_{-\infty}^0 0\mathrm{d}t = \frac{a+b}{2}.$$

If $b \le 0$, the expected value of ξ is

$$E[\xi] = \int_0^{+\infty} 0\mathrm{d}t - \left(\int_{-\infty}^{a} 0\mathrm{d}t + \int_a^b \frac{t-a}{b-a}\mathrm{d}t + \int_b^0 1\mathrm{d}t \right) = \frac{a+b}{2}.$$

If $a < 0 < b$, the expected value of ξ is

$$E[\xi] = \int_0^b \left(1 - \frac{t-a}{b-a}\right)\mathrm{d}t - \int_a^0 \frac{t-a}{b-a}\mathrm{d}t = \frac{a+b}{2}.$$

Therefore, we know that the expected value of ξ is always

$$E[\xi] = \frac{a+b}{2}.$$

Example 4.3. Let ξ be a normal uncertain variable $\mathcal{N}(e,\sigma)$. Then its expected value is

$$E[\xi] = e.$$

Theorem 4.9. *(Liu [64]) Let ξ be an uncertain variable whose uncertainty distribution is Φ. If its expected value exists, then*

$$E[\xi] = \int_0^1 \Phi^{-1}(\alpha)\mathrm{d}\alpha. \tag{4.16}$$

Proof: According to the definitions of expected value and uncertainty distribution, we have

$$\begin{aligned} E[\xi] &= \int_0^{+\infty} \mathcal{M}\{\xi \ge t\}\mathrm{d}t - \int_{-\infty}^0 \mathcal{M}\{\xi \le t\}\mathrm{d}t \\ &= \int_0^{+\infty} (1-\Phi(t))\mathrm{d}t - \int_{-\infty}^0 \Phi(t)\mathrm{d}t \\ &= \int_{\Phi(0)}^1 \Phi^{-1}(\alpha)\mathrm{d}\alpha + \int_0^{\Phi(0)} \Phi^{-1}(\alpha)\mathrm{d}\alpha = \int_0^1 \Phi^{-1}(\alpha)\mathrm{d}\alpha. \end{aligned}$$

Please also see Fig. 4.3.

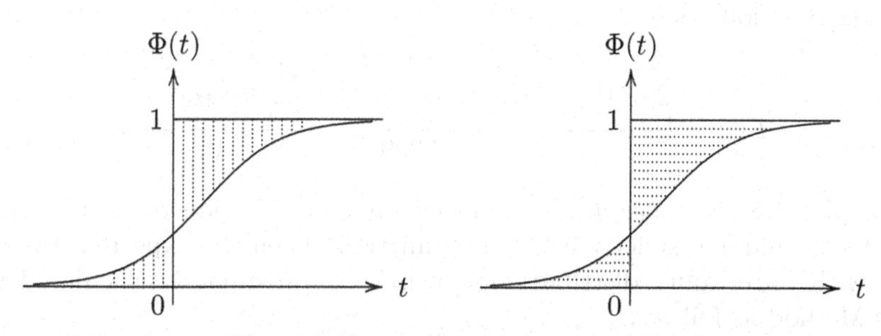

Fig. 4.3 Expected value via integral.

9999 Method B. *Suppose Φ is the uncertainty distribution of an uncertain variable ξ. Then ξ can be represented on a computer by*

α_i	0.0001	0.0002	0.0003	$\cdots$	0.9999
$\Phi^{-1}(\alpha_i)$	t_1	t_2	t_3	$\cdots$	t_{9999}

(4.17)

According to Theorem 4.9, the expected value of ξ can be approximately calculated by

$$E[\xi] = \frac{t_1 + t_2 + \cdots + t_{9999}}{9999}.$$

Remark 4.8. According to the precision requirement of the researcher, 9999 Method can also be a 99 Method or 999999 Method. For example, suppose Φ is the uncertainty distribution of an uncertain variable ξ. If the precision is required to be higher, then ξ can be represented on the computer by

α_i	0.000001	0.000002	0.000003	$\cdots$	0.999999
$\Phi^{-1}(\alpha_i)$	t_1	t_2	t_3	$\cdots$	t_{999999}

The expected value of ξ then can be approximately calculated via 999999 Method B as follows:

$$E[\xi] = \frac{t_1 + t_2 + \cdots + t_{999999}}{999999}.$$

Example 4.4. Suppose Φ_i is the uncertainty distribution of the uncertain variable ξ_i, and x_i a nonnegative number, $i = 1, 2, \cdots, n$, respectively. Then $\sum_{i=1}^{n} x_i \xi_i$ can be represented on a computer by

α_i	0.0001	0.0002	0.0003	$\cdots$	0.9999
$\Phi_i^{-1}(\alpha_i)$	$t_{i/1}$	$t_{i/2}$	$t_{i/3}$	$\cdots$	$t_{i/9999}$
$\Psi_i^{-1}(\alpha_i)$	$\sum_{i=1}^{n} x_i t_{i/1}$	$\sum_{i=1}^{n} x_i t_{i/2}$	$t_{i/3}$	$\cdots$	$\sum_{i=1}^{n} x_i t_{i/9999}$

The expected value of $\sum_{i=1}^{n} x_i \xi_i$ can be approximately calculated via 9999 Method B as follows:

$$E[\xi] = \frac{\sum_{i=1}^{n} x_1 t_{i/1} + \sum_{i=1}^{n} x_2 t_{i/2} + \cdots + \sum_{i=1}^{n} x_n t_{n/9999}}{9999}.$$

Example 4.5. Suppose Φ is the uncertainty distribution of an uncertain variable ξ, and f a strictly increasing function. Then the expected value of $f(\xi)$ with uncertainty distribution Ψ can be approximately calculated via 9999 Method as follows:

$$E[f(\xi)] = \frac{f(t_1) + f(t_2) + \cdots + f(t_{9999})}{9999}.$$

Theorem 4.10. *(Liu [64]) Let* ξ_1 *and* ξ_2 *be two independent uncertain variables with finite expected values. Then for any numbers* a_1 *and* a_2*, we have*

$$E[a_1\xi_1 + a_2\xi_2] = a_1E[\xi_1] + a_2E[\xi_2]. \tag{4.18}$$

Variance

Definition 4.9. *(Liu [60]) Let* ξ *be an uncertain variable with finite expected value* e*. Then the variance of* ξ *is defined by*

$$V[\xi] = E[(\xi - e)^2]. \tag{4.19}$$

Let ξ be an uncertain variable with uncertainty distribution Φ. Then

$$\begin{aligned}
V[\xi] &= \int_0^{+\infty} \mathcal{M}\{(\xi - e)^2\} \geq t\mathrm{d}t \\
&= \int_0^{+\infty} \mathcal{M}\{(\xi \geq e + \sqrt{t}) \cup (\xi \leq e - \sqrt{t})\}\mathrm{d}t \\
&\leq \int_0^{+\infty} (\mathcal{M}\{\xi \geq e + \sqrt{t}\} + \mathcal{M}\{\xi \leq e - \sqrt{t}\})\mathrm{d}t \\
&= \int_0^{+\infty} (1 - \Phi(e + \sqrt{t}) + \Phi(e - \sqrt{t}))\mathrm{d}t \\
&= \int_e^{+\infty} 2(t - e)(1 - \Phi(t) + \Phi(2e - t))\mathrm{d}t.
\end{aligned}$$

In this case, it is always assumed that the variance is

$$V[\xi] = 2\int_e^{+\infty} (t - e)(1 - \Phi(t) + \Phi(2e - t))\mathrm{d}t. \tag{4.20}$$

Example 4.6. Suppose ξ is a linear uncertain variable $(\mathcal{L}(a, b)$. Then its variance is

$$V[\xi] = 2\int_{(a+b)/2}^{b} \left(t - \frac{a+b}{2}\right)\left(1 - \frac{t-a}{b-a} + \frac{b-t}{b-a}\right)\mathrm{d}t = \frac{(b-a)^2}{12}.$$

Example 4.7. Suppose ξ is a normal uncertain variable $\mathcal{N}(e, \sigma)$. Then its variance is

$$V[\xi] = \sigma^2.$$

β-Value

Definition 4.10. *(Liu [60]) Let ξ be an uncertain variable, and $\beta \in (0, 1]$. Then*

$$\xi_{\sup}(\beta) = \sup\left\{r \mid \mathcal{M}\{\xi \geq r\} \geq \beta\right\} \tag{4.21}$$

is called the β-value to ξ.

Let ξ be an uncertain variable with continuous uncertainty distribution Φ such that

$$\lim_{t\to-\infty} \Phi(t) = 0, \quad \lim_{t\to\infty} \Phi(t) = 1.$$

Then its β-value is

$$\xi_{\sup}(\beta) = \Phi^{-1}(1-\beta)$$

provided that $\Phi^{-1}(\beta)$ is unique for each $\beta \in (0, 1]$ (see Fig. 4.4).

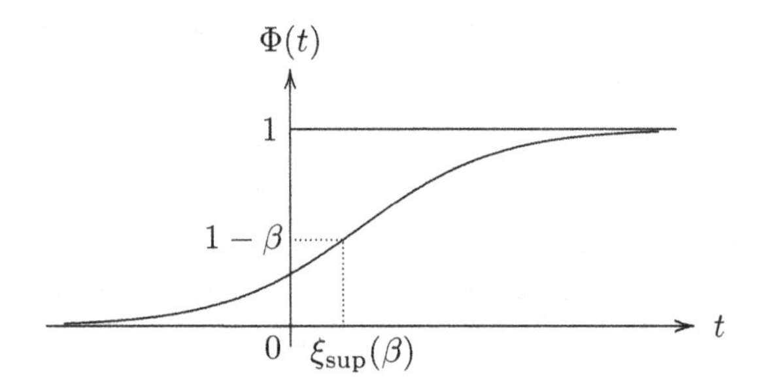

Fig. 4.4 β-Value.

Example 4.8. Let ξ be a linear uncertain variable $\mathcal{L}(a, b)$. Then its β-value is

$$\xi_{\sup}(\beta) = a\beta + b(1-\beta).$$

Example 4.9. Let ξ be a normal uncertain variable $\mathcal{N}(e, \sigma)$. Then its β-value is

$$\xi_{\sup}(\beta) = e - \frac{\sqrt{3}\sigma}{\pi} \ln \frac{\beta}{1-\beta}.$$

Theorem 4.11. *Let $\xi_{\sup}(\beta)$ be the β-value of an uncertain variable ξ. Then $\xi_{\sup}(\beta)$ is a decreasing and left-continuous function of β.*

Proof: Let β_1 and β_2 be two numbers with $0 < \beta_1 < \beta_2 \leq 1$. Then for any number $t < \xi_{\sup}(\beta_2)$, we have

$$\mathcal{M}\{\xi \geq t\} \geq \beta_2 > \beta_1.$$

Therefore, we obtain $\xi_{\sup}(\beta_1) \geq \xi_{\sup}(\beta_2)$ by the definition of β-value. That is, the β-value $\xi_{\sup}(\beta)$ is a decreasing function of β.

Now, we prove the left-continuity of $\xi_{\sup}(\beta)$ with respect to β. Let $\{\beta_i\}$ be an arbitrary sequence of positive numbers such that $\beta_i \uparrow \beta$. Then $\{\xi_{\sup}(\beta_i)\}$ is a decreasing sequence. If the limitation is equal to $\xi_{\sup}(\beta)$, then the left-continuity is proven. Otherwise, there exists a number z^* such that

$$\lim_{i\to\infty} \xi_{\sup}(\beta_i) > z^* > \xi_{\sup}(\beta).$$

Thus $\mathcal{M}\{\xi \geq z^*\} \geq \beta_i$ for each i. Letting $i \to \infty$, we get $M\{\xi \geq z^*\} \geq \beta$. Therefore, $z^* \leq \xi_{\sup}(\beta)$. A contradiction proves the left-continuity of $\xi_{\sup}(\beta)$ with respect to β.

Theorem 4.12. *Let $\xi_{\sup}(\beta)$ be the β-value of an uncertain variable ξ. Then if $\lambda \geq 0$, we have*

$$(\lambda\xi)_{\sup}(\beta) = \lambda\xi_{\sup}(\beta).$$

Proof: If $\lambda = 0$, the theorem holds obviously. If $\lambda > 0$, we have

$$\begin{aligned}(\lambda\xi)_{\sup}(\beta) &= \sup\{r|\mathcal{M}\{\lambda\xi \geq r\} \geq \beta\}\\ &= \lambda \sup\{r/\lambda|\mathcal{M}\{\xi \geq r/\lambda\} \geq \beta\}\\ &= \lambda\xi_{\sup}(\beta).\end{aligned}$$

Theorem 4.13. *Let ξ and η be two independent uncertain variables. Then for any $\beta \in (0,1]$, we have*

$$(\xi + \eta)_{\sup}(\beta) = \xi_{\sup}(\beta) + \eta_{\sup}(\beta). \tag{4.22}$$

Proof: According to monotonicity property of uncertain measure, for any $\epsilon > 0$, we have

$$\begin{aligned}&\mathcal{M}\{\xi + \eta \geq \xi_{\sup}(\beta) + \eta_{\sup}(\beta) - \epsilon\}\\ &\geq \mathcal{M}\Big\{\{\xi \geq \xi_{\sup}(\beta) - \epsilon/2\} \cap \{\eta \geq \eta_{\sup}(\beta) - \epsilon/2\}\Big\}.\end{aligned}$$

Since ξ and η are independent uncertain variables, according to Equation (4.4), we have

$$\begin{aligned}&\mathcal{M}\{\xi + \eta \geq \xi_{\sup}(\beta) + \eta_{\sup}(\beta) - \epsilon\}\\ &\geq \mathcal{M}\Big\{\{\xi \geq \xi_{\sup}(\beta) - \epsilon/2\} \cap \{\eta \geq \eta_{\sup}(\beta) - \epsilon/2\}\Big\}\\ &= \mathcal{M}\{\xi \geq \xi_{\sup}(\beta) - \epsilon/2\} \wedge \mathcal{M}\{\eta \geq \eta_{\sup}(\beta) - \epsilon/2\} \geq \beta\end{aligned}$$

which implies that

$$(\xi + \eta)_{\sup}(\beta) \geq \xi_{\sup}(\beta) + \eta_{\sup}(\beta) - \epsilon. \tag{4.23}$$

According to monotonicity property of uncertain measure, for any $\epsilon > 0$, we have

$$\begin{aligned}&\mathcal{M}\{\xi+\eta \geq \xi_{\sup}(\beta)+\eta_{\sup}(\beta)+\epsilon\}\\ &\leq \mathcal{M}\Big\{\{\xi \geq \xi_{\sup}(\beta)+\epsilon/2\} \cup \{\eta \geq \eta_{\sup}(\beta)+\epsilon/2\}\Big\}.\end{aligned}$$

Since ξ and η are independent uncertain variables, according to Equation (4.5), we have

$$\begin{aligned}&\mathcal{M}\{\xi+\eta \geq \xi_{\sup}(\beta)+\eta_{\sup}(\beta)+\epsilon\}\\ &\leq \mathcal{M}\Big\{\{\xi \geq \xi_{\sup}(\beta)+\epsilon/2\} \cap \{\eta \geq \eta_{\sup}(\beta)+\epsilon/2\}\Big\}\\ &= \mathcal{M}\{\xi \geq \xi_{\sup}(\beta)+\epsilon/2\} \vee \mathrm{Cr}\{\eta \geq \eta_{\sup}(\beta)+\epsilon/2\} < \beta\end{aligned}$$

which implies that

$$(\xi+\eta)_{\sup}(\beta) \leq \xi_{\sup}(\beta)+\eta_{\sup}(\beta)+\epsilon. \tag{4.24}$$

It follows from (4.23) and (4.24) that

$$\xi_{\sup}(\beta)+\eta_{\sup}(\beta)+\epsilon \geq (\xi+\eta)_{\sup}(\beta) \geq \xi_{\sup}(\beta)+\eta_{\sup}(\beta)-\epsilon.$$

Letting $\epsilon \rightarrow 0$, we have

$$(\xi+\eta)_{\sup}(\beta) = \xi_{\sup}(\beta)+\eta_{\sup}(\beta).$$

4.2 Mean-Risk Model

Following the idea of risk curve in probabilistic and credibilistic portfolio selection, Huang [42] proposed a concept of risk curve to give an instinct information about each likely loss and the corresponding occurrence chance of the loss for portfolio selection problem in which the security returns are believed to be uncertain variables.

4.2.1 Risk Curve

Definition 4.11. *(Huang [42]) Let ξ denote an uncertain return of a portfolio, and r_f the risk-free interest rate. Then the curve*

$$R(r) = \mathcal{M}\{r_f - \xi \geq r\}, \quad \forall r \geq 0 \tag{4.25}$$

is called the risk curve of the portfolio, and r the loss severity indicator.

It is clear that $r_f - \xi$ is the value that the portfolio return is lower than the risk-free interest rate when $r_f - \xi \geq 0$. This value $r_f - \xi$ can certainly be understood as a loss. Since the portfolio return is variable, the loss level $r_f - \xi$ may be any non-negative values which can be expressed by

$$r_f - \xi \geq r, \quad r \geq 0.$$

Since r is not one specific number but any non-negative numbers, the risk curve $R(r)$ describes all the likely losses of the portfolio and the corresponding occurrence chances of all these losses.

According to the monotonicity axiom of the uncertain measure, the risk curve $R(r)$ is a decreasing function of the real numbers r. That is, when the loss becomes bigger, the occurrence chance of the loss will become smaller.

Equivalently, the risk curve can also be expressed in the form

$$R^{-1}(\alpha) = r_f - \Phi^{-1}(\alpha), \quad \forall \alpha \in (0,1) \tag{4.26}$$

where Φ is the uncertainty distribution of ξ.

With Formulation (4.26), given a confidence level, the investors are able to know how much they will lose at this occurrence chance level.

Example 4.10. If a portfolio return is a linear uncertain variable $\xi \sim \mathcal{L}(a, b)$, the risk curve of the portfolio is as follows (also see Fig. 4.5):

$$R(r) = \mathcal{M}\{(r_f - \xi) \geq r\} = \begin{cases} 1, & \text{if } r < r_f - b \\ \dfrac{r_f - a - r}{b - a}, & \text{if } r_f - b \leq r \leq r_f - a \\ 0, & \text{if } r > r_f - a. \end{cases}$$

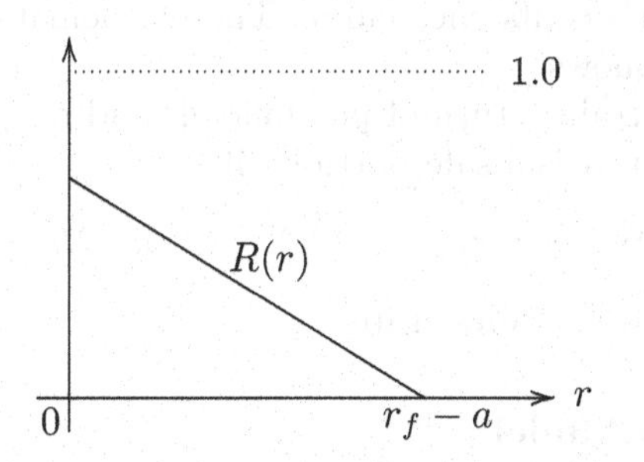

Fig. 4.5 Risk curve of a portfolio with linear uncertain return.

Example 4.11. If a portfolio return is a normal uncertain variable $\xi \sim \mathcal{N}(e, \sigma)$. Then the risk curve of the portfolio is as follows (also see Fig. 4.6):

$$R(r) = \mathcal{M}\{r_f - \xi \geq r\} = \left(1 + \exp\left(\frac{\pi(e - r_f + r)}{\sqrt{3}\sigma}\right)\right)^{-1}, \quad \forall r \geq 0.$$

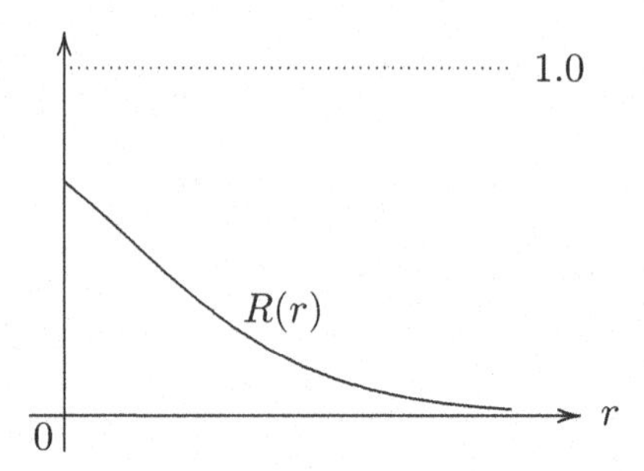

Fig. 4.6 Risk curve of a portfolio with normal uncertain return.

4.2.2 Confidence Curve and Safe Portfolio

Since all investors know that they may lose as well as gain in investment, they will have a maximum tolerance towards occurrence chance of each likely loss level. We call the curve confidence curve $\alpha(r)$ that gives the investors' maximal tolerance towards the occurrence chance of each likely loss level. An investor can give his/her confidence curve by answering what-if questions in Table 2.5. In uncertain portfolio selection, occurrence chance of an uncertain event is measured by uncertain measure. Though different investors have different confidence curve, the common trend of the curve is that the severer the loss, the lower the loss occurrence chance the investors can tolerate. That is, the higher the r value, the lower the uncertain measure value. Three examples of confidence curve are given in Subsection 2.2.2.

It is clear that a portfolio is safe if its risk curve is totally below the investor's confidence curve. A portfolio is risky if any part of its risk curve is above the investor's confidence curve. The mathematical expression for a safe portfolio is as follows:

Let ξ be the uncertain return of portfolio A, and $\alpha(r)$ the investor's confidence curve. We say A is a safe portfolio if

$$R(r) = \mathcal{M}\{(r_f - \xi) \geq r\} \leq \alpha(r), \quad \forall r \geq 0,$$

where r_f is the risk-free interest rate.

4.2.3 Mean-Risk Model

The philosophy of mean-risk model is to pursue maximum expected return among the safe portfolios whose risk curves are below the investor's confidence curve. Let x_i denote the investment proportions in securities $i, i = 1, 2, \cdots, n$, respectively, and ξ_i the i-th security returns which are uncertain variables. According to the risk definition (4.25), the risk curve of a portfolio $(x_1, x_2, \cdots, x_n)$ is

$$R(x_1, x_2, \cdots, x_n; r) = \mathcal{M}\left\{r_f - (\xi_1 x_1 + \xi_2 x_2 + \cdots + \xi_n x_n) \geq r\right\}. \quad (4.27)$$

Let $\alpha(r)$ be an investor's confidence curve. The mean-risk model for portfolio selection with uncertain returns is expressed as follows:

$$\begin{cases} \max E[\xi_1 x_1 + \xi_2 x_2 + \cdots + \xi_n x_n] \\ \text{subject to:} \\ \qquad R(x_1, x_2, \cdots, x_n; r) \le \alpha(r), \ \forall r \ge 0 \\ \qquad x_1 + x_2 + \cdots + x_n = 1 \\ \qquad x_i \ge 0, \quad i = 1, 2, \cdots, n \end{cases} \tag{4.28}$$

where E is the expected value operator, $R(x_1, x_2, \cdots, x_n; r)$ the risk curve defined by Equation (4.27), and $\alpha(r)$ the investor's confidence curve. The constraint

$$R(x_1, x_2, \cdots, x_n; r) \le \alpha(r), \ \forall r \ge 0$$

means for any given loss level r, the occurrence chance of the loss is not greater than the investor's tolerable occurrence chance level, which ensures that the optimal portfolio is selected among the safe portfolios (see Fig. 4.7).

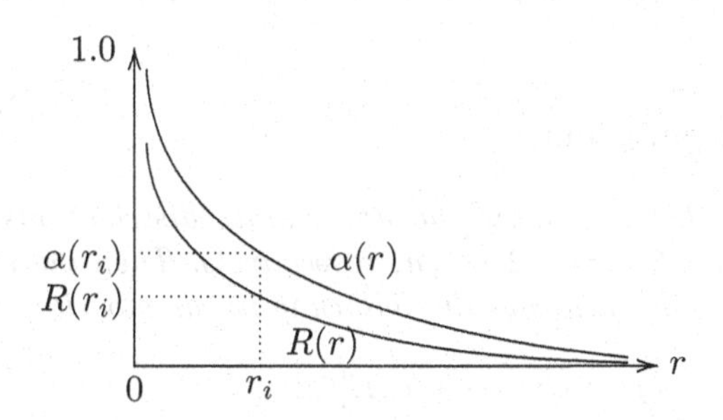

Fig. 4.7 Safe portfolio: A portfolio is safe if $R(r) \le \alpha(r)$ for any $r \ge 0$.

According to Formula (4.26), the mean-risk model (4.28) can also be expressed in the following way:

$$\begin{cases} \max E[\xi_1 x_1 + \xi_2 x_2 + \cdots + \xi_n x_n] \\ \text{subject to:} \\ \qquad R^{-1}(x_1, x_2, \cdots, x_n; r)(\alpha(r)) \le r, \ \forall \alpha(r) \ge 0 \\ \qquad x_1 + x_2 + \cdots + x_n = 1 \\ \qquad x_i \ge 0, \quad i = 1, 2, \cdots, n \end{cases} \tag{4.29}$$

where

$$R^{-1}(x_1, x_2, \cdots, x_n; r)(\alpha(r)) = r_f - \Psi(\alpha(r))$$

in which Ψ is the uncertainty distribution of the uncertain variable $\xi_1 x_1 + \xi_2 x_2 + \cdots + \xi_n x_n$. The constraint

$$R^{-1}(x_1, x_2, \cdots, x_n; r)(\alpha(r)) \leq r, \ \forall \alpha(r) \geq 0$$

means that given any confidence level, the loss level occurring at the confidence level is not greater than the investor's tolerable level, which ensures that the optimal portfolio is selected among the safe portfolios (see Fig. 4.8).

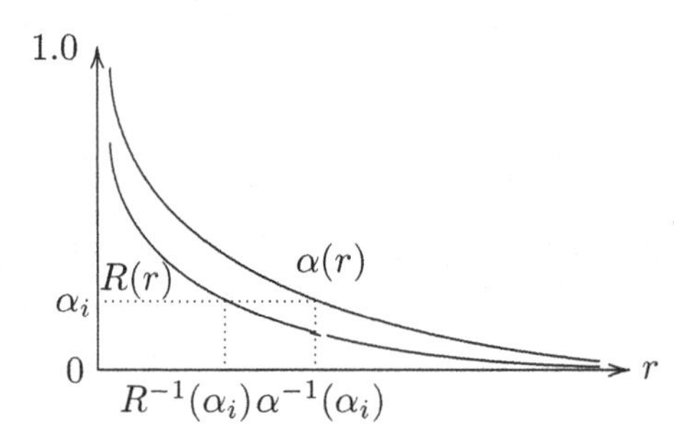

Fig. 4.8 Safe portfolio: A portfolio is safe if $R^{-1}(\alpha(r)) \leq r$ for any confidence $\alpha(r) \geq 0$.

4.2.4 Crisp Equivalent

Theorem 4.14. *Let Φ_i denote the uncertainty distributions of the i-th security return rates $\xi_i, i = 1, 2, \cdots, n$, respectively. Then the mean-risk model (4.28) can be transformed into the following linear model:*

$$\begin{cases} \max x_1 E[\xi_1] + x_2 E[\xi_2] + \cdots + x_n E[\xi_n] \\ \textit{subject to:} \\ \qquad x_1 \Phi_1^{-1}\Big(\alpha(r)\Big) + x_2 \Phi_2^{-1}\Big(\alpha(r)\Big) + \cdots + x_n \Phi_n^{-1}\Big(\alpha(r)\Big) \geq r_f - r, \ \forall r \geq 0 \\ \qquad x_1 + x_2 + \cdots + x_n = 1 \\ \qquad x_i \geq 0, \quad i = 1, 2, \cdots, n. \end{cases} \tag{4.30}$$

The objective can be obtained by the linearity property of the expected value of the uncertain variables, and the constraint can be proven directly from Theorem 4.6 and the monotonicity property of the uncertain measure.

Example 4.12. Suppose the return rates of the i-th securities are all normal uncertain variables $\xi_i \sim \mathcal{N}(e_i, \sigma_i), i = 1, 2, \cdots, n$, respectively. Then the mean-risk model can be transformed into the following form:

$$
\left\{
\begin{array}{l}
\max e_1x_1 + e_2x_2 + \cdots + e_nx_n \\
\text{subject to:} \\
\qquad \sum_{i=1}^{n} \left(e_i - \frac{\sqrt{3}\sigma_i}{\pi} \ln \frac{1-\alpha(r)}{\alpha(r)} \right) x_i \geq r_f - r, \quad \forall r \geq 0 \\
\qquad x_1 + x_2 + \cdots + x_n = 1 \\
\qquad x_i \geq 0, \quad i = 1, 2, \cdots, n.
\end{array}
\right.
\tag{4.31}
$$

Example 4.13. Suppose the return rates of the i-th securities are all linear uncertain variables $\xi_i = \mathcal{L}(a_i, b_i), i = 1, 2, \cdots, n$, respectively. Then the mean-risk model can be transformed into the following form:

$$
\left\{
\begin{array}{l}
\max \sum_{i=1}^{n} \frac{1}{2}(b_i + a_i)x_i \\
\text{subject to:} \\
\qquad \sum_{i=1}^{n} \alpha(r)b_ix_i + \sum_{i=1}^{n} \Big(1 - \alpha(r)\Big) a_ix_i \geq r_f - r, \quad r \geq 0 \\
\qquad x_1 + x_2 + \cdots + x_n = 1 \\
\qquad x_i \geq 0, \quad i = 1, 2, \cdots, n.
\end{array}
\right.
\tag{4.32}
$$

Example 4.14. Suppose the return rates of the i-th securities are normal uncertain variables $\xi_i \sim \mathcal{N}(e_i, \sigma_i), i = 1, 2, \cdots, m$, and the return rates of the j-th securities are all linear uncertain variables $\xi_j = \mathcal{L}(a_j, b_j), j = m+1, m+2, \cdots, n$, respectively. Then the mean-risk model can be transformed into the following form:

$$
\left\{
\begin{array}{l}
\max \sum_{i=1}^{m} e_ix_i + \sum_{i=m+1}^{n} \frac{1}{2}(b_i + a_i)x_i \\
\text{subject to:} \\
\qquad \sum_{i=1}^{m} \left(e_i - \frac{\sqrt{3}\sigma_i}{\pi} \ln \frac{1-\alpha(r)}{\alpha(r)} \right) x_i + \sum_{i=m+1}^{n} \alpha(r)b_ix_i \\
\qquad + \sum_{i=m+1}^{n} \Big(1 - \alpha(r)\Big) a_ix_i \geq r_f - r, \quad r \geq 0 \\
\qquad x_1 + x_2 + \cdots + x_n = 1 \\
\qquad x_i \geq 0, \quad i = 1, 2, \cdots, n.
\end{array}
\right.
\tag{4.33}
$$

4.2.5 Examples

Example 4.15. Suppose an investor chooses from 10 securities for his/her investment. Assume that the monthly return rates of the ten securities are all normal uncertain variables denoted by $\xi_i, i = 1, 2, \cdots, 10$, respectively. The prediction of the return rates of the ten securities is given in Table 4.1.

Table 4.1 Uncertain Return Rates of 10 Securities

Security i	Return Rate ξ_i	Security i	Return Rate ξ_i
1	$\mathcal{N}(0.027, 0.14)$	6	$\mathcal{N}(0.028, 0.15)$
2	$\mathcal{N}(0.033, 0.12)$	7	$\mathcal{N}(0.030, 0.12)$
3	$\mathcal{N}(0.032, 0.16)$	8	$\mathcal{N}(0.032, 0.18)$
4	$\mathcal{N}(0.044, 0.16)$	9	$\mathcal{N}(0.025, 0.10)$
5	$\mathcal{N}(0.031, 0.15)$	10	$\mathcal{N}(0.028, 0.11)$

Suppose the monthly risk-free interest rate is 0.01, and the investor gives his/her confidence curve as follows:

$$\alpha(r) = \begin{cases} -2.75r + 0.43, & 0 \le r \le 0.12, \\ -0.5r + 0.16, & 0.12 \le r \le 0.3, \\ 0.01, & r \ge 0.3. \end{cases}$$

According to the mean-risk selection idea, we build the model as follows:

$$\begin{cases} \max E[\xi_1 x_1 + \xi_2 x_2 + \cdots + \xi_{10} x_{10}] \\ \text{subject to:} \\ \qquad R(x_1, x_2, \cdots, x_{10}; r) \le \alpha(r), \ \forall r \ge 0 \\ \qquad x_1 + x_2 + \cdots + x_{10} = 1 \\ \qquad x_i \ge 0, \quad i = 1, 2, \cdots, 10 \end{cases} \tag{4.34}$$

where $R(x_1, x_2, \cdots, x_{10}; r)$ is the risk curve of the portfolio and

$$R(x_1, x_2, \cdots, x_{10}; r) = \mathcal{M}\{0.01 - (\xi_1 x_1 + \xi_2 x_2 + \cdots + \xi_{10} x_{10}) \ge r\}.$$

It follows from Model (4.31) that Model (4.34) can be transformed into the following form:

$$\begin{cases} \max \sum_{i=1}^{10} e_i x_i \\ \text{subject to:} \\ \quad \sum_{i=1}^{10} \left(e_i - \frac{\sqrt{3}\sigma_i}{\pi} \ln \frac{1-\alpha(r)}{\alpha(r)} \right) x_i \geq 0.01 - r, \quad r \geq 0 \\ \quad x_1 + x_2 + \cdots + x_{10} = 1 \\ \quad x_i \geq 0, \quad i = 1, 2, \cdots, 10. \end{cases} \tag{4.35}$$

Model (4.35) is a linear programming model which can be solved by the simplex method. Though theoretically, r should be any non-negative numbers, in reality, we just need to check it in a certain interval by analyzing the problem and the confidence curve. In the example, the confidence curve is a horizontal line when $r \geq 0.3$. Since the risk curve is a decreasing function of r, risk curve will be below the confidence curve if $R(x_1, x_2, \cdots, x_{10}; r) \leq \alpha(r)$ holds for any $r \in [0, 0.3]$. Since risk curve is a continuous function of r, it is enough for us to check if the points on the risk curve are all lower than the points on the confidence curve for $(r = 0, \alpha = 0.43)$, $(r = 0.03, \alpha = 0.3475)$ $(r = 0.06, \alpha = 0.265) \cdots$, $(r = 0.3, \alpha = 0.01)$. That is, we just need to solve Model (4.36) given below.

$$\begin{cases} \max \sum_{i=1}^{10} e_i x_i \\ \text{subject to:} \\ \quad \sum_{i=1}^{10} \left(e_i - \frac{\sqrt{3}\sigma_i}{\pi} \ln \frac{1-0.43}{0.43} \right) x_i \geq 0.01 \\ \quad \sum_{i=1}^{10} \left(e_i - \frac{\sqrt{3}\sigma_i}{\pi} \ln \frac{1-0.3475}{0.3475} \right) x_i \geq -0.02 \\ \quad \sum_{i=1}^{10} \left(e_i - \frac{\sqrt{3}\sigma_i}{\pi} \ln \frac{1-0.265}{0.265} \right) x_i \geq -0.05 \\ \quad \cdots \\ \quad \sum_{i=1}^{10} \left(e_i - \frac{\sqrt{3}\sigma_i}{\pi} \ln \frac{1-0.01}{0.01} \right) x_i \geq -0.31 \\ \quad x_1 + x_2 + \cdots + x_{10} = 1 \\ \quad x_i \geq 0, \quad i = 1, 2, \cdots, 10. \end{cases} \tag{4.36}$$

The result shows that to gain the maximal expected return among the safe portfolios, the investor should allocate his/her money according to Table 4.2.

Table 4.2 Allocation of Money to Ten Securities

Security i	1	2	3	4	5
Allocation of money	0.00%	85.41%	0.00%	0.00%	0.00%
Security i	6	7	8	9	10
Allocation of money	0.00%	0.00%	0.00%	14.59%	0.00%

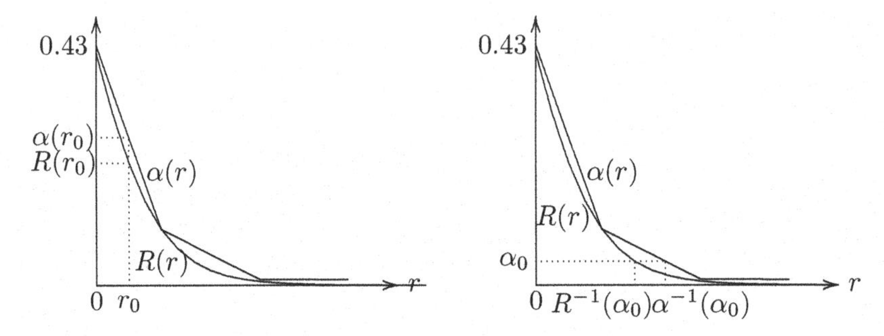

Fig. 4.9 Risk curve $R(r)$ and confidence curve $\alpha(r)$ of Model (4.34).

The maximal expected return rate is 0.032. It can be seen from Fig. 4.9 that the risk cure of the selected portfolio $R(x_1, x_2, \cdots, x_{10}; r)$ is under the investor's confidence curve $\alpha(r)$. Given a loss level $r = r_0$, the occurrence chance of the loss $R(r_0)$ is below the investor's tolerable occurrence chance level $\alpha(r_0)$; or given a confidence level $\alpha = \alpha_0$, the loss level $R^{-1}(\alpha_0)$ is less than the investor's tolerable loss level $\alpha^{-1}(\alpha_0)$.

Example 4.16. Suppose now the investor chooses an optimal portfolio from another ten securities of which seven security return rates are normal uncertain variables and the rest three the linear uncertain variables. The prediction of the return rates of the 10 securities is given in Table 4.3. The investor's confidence curve and the risk-free interest rate are the same as the above mentioned example. Then the optimal portfolio can be selected according to the following model:

$$\begin{cases} \max \sum_{i=1}^{7} e_i x_i + \sum_{i=8}^{10} \frac{b_i + a_i}{2} x_i \\ \text{subject to:} \\ \qquad \sum_{i=1}^{7} \left(e_i - \frac{\sqrt{3}\sigma_i}{\pi} \ln \frac{1-\alpha(r)}{\alpha(r)} \right) x_i + \sum_{i=8}^{10} \alpha(r) b_i x_i \\ \qquad + \sum_{i=8}^{10} \Big(1 - \alpha(r)\Big) a_i x_i \geq 0.01 - r, \quad r \geq 0 \\ \qquad x_1 + x_2 + \cdots + x_{10} = 1 \\ \qquad x_i \geq 0, \quad i = 1, 2, \cdots, 10. \end{cases} \tag{4.37}$$

Table 4.3 Uncertain Return Rates of 10 Securities

Security i	Return Rate ξ_i	Security i	Return Rate ξ_i
1	$\mathcal{N}(0.033, 0.19)$	6	$\mathcal{N}(0.026, 0.06)$
2	$\mathcal{N}(0.032, 0.16)$	7	$\mathcal{N}(0.030, 0.08)$
3	$\mathcal{N}(0.039, 0.20)$	8	$\mathcal{L}(-0.1, 0.16)$
4	$\mathcal{N}(0.031, 0.15)$	9	$\mathcal{L}(-0.15, 0.22)$
5	$\mathcal{N}(0.025, 0.10)$	10	$\mathcal{L}(-0.2, 0.3)$

Table 4.4 Allocation of Money to Ten Securities

Security i	1	2	3	4	5
Allocation of money	0.00%	0.00%	0.00%	0.00%	0.00%
Security i	6	7	8	9	10
Allocation of money	38.72%	0.00%	0.00%	0.00%	61.28%

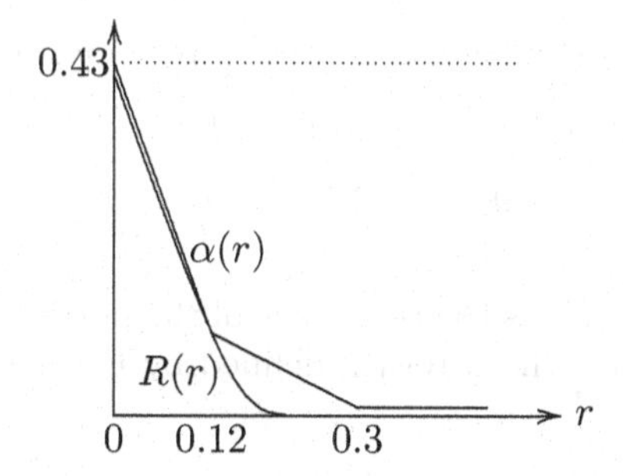

Fig. 4.10 Risk curve $R(r)$ and confidence curve $\alpha(r)$ of Model (4.37).

By running "Solver" in "Excel", we get the result that to gain the maximal expected return among the safe portfolios, the investor should allocate his/her money according to Table 4.4. The maximal expected return rate is 0.0407. It can be seen from Fig. 4.10 that the risk cure of the selected portfolio $R(x_1, x_2, \cdots, x_{10}; r)$ is under the investor's confidence curve $\alpha(r)$.

4.3 β-Return-Risk Model

4.3.1 β-Return-Risk Model

Risk curve provides instinct information about likely losses and the corresponding occurrence chance levels. To provide an instinct information about return, β-Return is proposed.

Definition 4.12. *Let x_i be the investment proportions in the i-th securities, $i = 1, 2, \cdots, n$, respectively, ξ_i the uncertain returns of the i-th securities and β the preset confidence level. The β-return is defined as*

$$\max\{\bar{f} \mid \mathcal{M}\{\xi_1 x_1 + \xi_2 x_2 + \cdots + \xi_n x_n \geq \bar{f}\} \geq \beta\} \tag{4.38}$$

which means the maximal investment return the investor can obtain at confidence level β.

When the investors want to pursue an instinct and a specific maximum return rather than an expected value, they can ask that the risk curve of the portfolio be totally below the confidence curve, and in the meantime pursue the maximum return they can obtain at a high enough occurrence chance level β. To express the idea mathematically, we have the β-return-risk model as follows:

$$\begin{cases} \max \bar{f} \\ \text{subject to:} \\ \qquad \mathcal{M}\{\xi_1 x_1 + \xi_2 x_2 + \cdots + \xi_n x_n \geq \bar{f}\} \geq \beta \\ \qquad R(x_1, x_2, \cdots, x_n; r) \leq \alpha(r), \forall r \geq 0 \\ \qquad x_1 + x_2 + \cdots + x_n = 1 \\ \qquad x_i \geq 0, \quad i = 1, 2, \cdots, n \end{cases} \tag{4.39}$$

where $R(x_1, x_2, \cdots, x_n; r)$ is the risk curve of the portfolio, $\alpha(r)$ the investors' confidence curve, and $\bar{f}$ the β-return defined by Formula (4.38).

4.3.2 Crisp Equivalent

When security returns are independent uncertain variables, we can change the β-return-risk model into its equivalent and solve the model in traditional ways.

Theorem 4.15. *Let Φ_i denote the uncertainty distributions of the i-th security return rates $\xi_i, i = 1, 2, \cdots, n$, respectively. Then the β-return-risk model (4.39) can be transformed into the following linear model:*

$$\begin{cases} \max x_1 \xi_1(\beta) + +x_2 \xi_2(\beta) + \cdots + x_n \xi_n(\beta) \\ \textit{subject to:} \\ \qquad x_1 \Phi_1^{-1}\Big(\alpha(r)\Big) + x_2 \Phi_2^{-1}\Big(\alpha(r)\Big) + \cdots + x_n \Phi_n^{-1}\Big(\alpha(r)\Big) \geq r_f - r, \ \forall r \geq 0 \\ \qquad x_1 + x_2 + \cdots + x_n = 1 \\ \qquad x_i \geq 0, \quad i = 1, 2, \cdots, n \end{cases} \tag{4.40}$$

where $\xi_i(\beta)$ is the β-return of the i-th security which is defined by

$$\max\{\bar{f}_i \mid \mathcal{M}\{\xi_i \geq \bar{f}_i\} \geq \beta\}$$

The objective function can be obtained directly from Theorem 4.13.

Example 4.17. Suppose the return rates of the i-th securities are all normal uncertain variables $\xi_i \sim \mathcal{N}(e_i, \sigma_i), i = 1, 2, \cdots, n$, respectively. Then the β-return-risk model can be transformed into the following form:

$$\begin{cases} \max \displaystyle\sum_{i=1}^{n} e_i x_i - \frac{\sqrt{3}}{\pi} \ln \frac{\beta}{1-\beta} \sum_{i=1}^{n} \sigma_i x_i \\ \text{subject to:} \\ \qquad \displaystyle\sum_{i=1}^{n} \left(e_i - \frac{\sqrt{3}\sigma_i}{\pi} \ln \frac{1-\alpha(r)}{\alpha(r)} \right) x_i \geq r_f - r, \quad \forall r \geq 0 \\ \qquad x_1 + x_2 + \cdots + x_n = 1 \\ \qquad x_i \geq 0, \quad i = 1, 2, \cdots, n. \end{cases} \tag{4.41}$$

Since the weighted sum of normal uncertain variables is still a normal uncertain variable, the objective function of Model (4.41) can be easily obtained from Theorem 4.15.

Example 4.18. Suppose the return rates of the i-th securities are all linear uncertain variables $\xi_i = \mathcal{L}(a_i, b_i), i = 1, 2, \cdots, n$, respectively. Then the β-return-risk model can be transformed into the following form:

$$\begin{cases} \max \displaystyle\sum_{i=1}^{n} \beta a_i x_i + (1-\beta) \sum_{i=1}^{n} b_i x_i \\ \text{subject to:} \\ \qquad \displaystyle\sum_{i=1}^{n} \alpha(r) b_i x_i + \sum_{i=1}^{n} \Big(1 - \alpha(r)\Big) a_i x_i \geq r_f - r, \quad r \geq 0 \\ \qquad x_1 + x_2 + \cdots + x_n = 1 \\ \qquad x_i \geq 0, \quad i = 1, 2, \cdots, n. \end{cases} \tag{4.42}$$

Since the weighted sum of linear uncertain variables is still a linear uncertain variable, the objective function of Model (4.42) can be easily obtained from Theorem 4.15.

Example 4.19. Suppose the return rates of the i-th securities are normal uncertain variables $\xi_i \sim \mathcal{N}(e_i, \sigma_i), i = 1, 2, \cdots, m$, and the return rates of the j-th securities are all linear uncertain variables $\xi_j = \mathcal{L}(a_j, b_j), j = m+1, m+$

$2, \cdots, n$, respectively. Then the β-return-risk model can be transformed into the following form:

$$\begin{cases} \max \sum_{i=1}^{m} e_i x_i - \frac{\sqrt{3}}{\pi} \ln \frac{\beta}{1-\beta} \sum_{i=1}^{m} \sigma_i x_i + \beta \sum_{i=1}^{n} a_i x_i + (1-\beta) \sum_{i=1}^{n} b_i x_i \\ \text{subject to:} \\ \qquad \sum_{i=1}^{m} \left(e_i - \frac{\sqrt{3}\sigma_i}{\pi} \ln \frac{1-\alpha(r)}{\alpha(r)} \right) x_i + \sum_{i=m+1}^{n} \alpha(r) b_i x_i \\ \qquad + \Big(1-\alpha(r)\Big) a_i x_i \geq r_f - r, \quad r \geq 0 \\ \qquad x_1 + x_2 + \cdots + x_n = 1 \\ \qquad x_i \geq 0, \quad i = 1, 2, \cdots, n. \end{cases} \tag{4.43}$$

4.3.3 An Example

Example 4.20. Recall the 10 securities whose returns are given in Table 4.1 in subsection 4.2.5. The risk-free return rate is 0.01, and the investor's confidence curve is

$$\alpha(r) = \begin{cases} -2.75r + 0.43, & 0 \leq r \leq 0.12, \\ -0.5r + 0.16, & 0.12 \leq r \leq 0.3, \\ 0.01, & r \geq 0.3. \end{cases}$$

Suppose this time the investor wants to pursue a maximum specific return at confidence 80% from the safe portfolios. Then he/she should select the portfolio based on the following model:

$$\begin{cases} \max \xi_1(0.80)x_1 + \xi_2(0.80)x_2 + \cdots + \xi_{10}(0.80)x_{10} \\ \text{subject to:} \\ \qquad R(x_1, x_2, \cdots, x_{10}; r) \leq \alpha(r), \ \forall r \geq 0 \\ \qquad x_1 + x_2 + \cdots + x_{10} = 1 \\ \qquad x_i \geq 0, \quad i = 1, 2, \cdots, 10 \end{cases} \tag{4.44}$$

where $\xi_i(0.80)$ is the 0.80-return of the i-th security which is defined as

$$\xi_i(0.80) = \sup\{\bar{f}_i | \mathcal{M}\{\xi_i \geq \bar{f}_i\} \geq 0.80\},$$

and $R(x_1, x_2, \cdots, x_{10}; r)$ the risk curve of the portfolio defined as

$$R(x_1, x_2, \cdots, x_{10}; r) = \mathcal{M}\left\{0.01 - (\xi_1 x_1 + \xi_2 x_2 + \cdots + \xi_{10} x_{10}) \geq r\right\}.$$

Table 4.5 Allocation of Money to Ten Securities

Security i	1	2	3	4	5
Allocation of money	0.00%	11.02%	0.00%	0.00%	0.00%
Security i	6	7	8	9	10
Allocation of money	0.00%	0.00%	0.00%	88.98%	0.00%

It follows from Model (4.41) that Model (4.44) can be transformed into the following form:

$$\begin{cases} \max \sum_{i=1}^{10} e_i x_i - \frac{\sqrt{3}}{\pi} \ln 4 \sum_{i=1}^{10} \sigma_i x_i \\ \text{subject to:} \\ \quad \sum_{i=1}^{10} \left(e_i - \frac{\sqrt{3}\sigma_i}{\pi} \ln \frac{1-\alpha(r)}{\alpha(r)} \right) x_i \geq 0.01 - r, \quad r \geq 0 \\ \quad x_1 + x_2 + \cdots + x_{10} = 1 \\ \quad x_i \geq 0, \quad i = 1, 2, \cdots, 10. \end{cases} \tag{4.45}$$

By running "Solver" in "Excel" we get the result that to gain the maximal 0.8-return among the safe portfolios, the investor should allocate his/her money according to Table 4.5. The maximal 0.8-return rate is -0.052 which means that the investor will gain no less than -0.052 (or say lose no more than 0.052 below zero) at confidence 0.8. It can be seen from Fig. 4.11 that the risk cure of the selected portfolio $R(x_1, x_2, \cdots, x_{10}; r)$ is under the investor's confidence curve $\alpha(r)$.

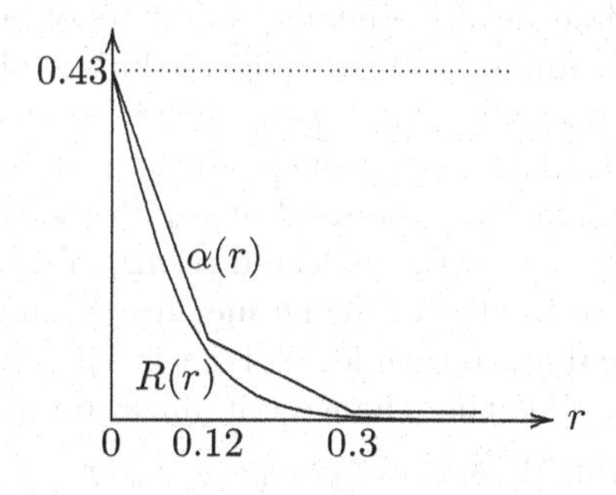

Fig. 4.11 Risk curve $R(r)$ and confidence curve $\alpha(r)$ of Model (4.44).

Remark 4.9. We put the results of examples of mean-risk model in subsection 4.2.5 and β-return-risk model in this subsection together in Table 4.6. It is seen that even when risk-free interest rate, the alternative individual

Table 4.6 Optimal Portfolios Produced by Different Selection Criteria

Optimal Portfolio	Mean-Risk Criterion	β-Return-Risk Criterion
ξ_1	0.00%	0.00%
ξ_2	85.41%	11.02%
ξ_3	0.00%	0.00%
ξ_4	0.00%	0.00%
ξ_5	0.00 %	0.00%
ξ_6	0.00 %	0.00%
ξ_7	0.00%	0.00%
ξ_8	0.00%	0.00%
ξ_9	14.59%	88.98%
ξ_{10}	0.00%	0.00%
Expected Return	3.2%	2.6%
0.8-Return	−11%	−5.2%

securities and the investor's confidence curve are same, adopting different selection criteria produces different results.

4.4 Chance Minimization Model

4.4.1 Chance Minimization Model

Risk curve provides a panoramic view of all the likely losses of an investment. The investors who adopt the risk curve as the investment risk are the most cautious investors. They evaluate every likely loss case and compare it with their own tolerance ability. Therefore, the decision making based on mean-risk or β-return-risk model is the safest. However, to use mean-risk or β-return-risk model, the investors need to provide their confidence curve and find out occurrence chances of big enough numbers of loss cases. Sometimes, the investors want a simpler way to judge riskiness of an investment and to select an optimal portfolio. They may only be sensitive to one very low return event (or say one large loss event). Then the occurrence chance of the concerned low return event can be used as an alternative definition of risk. The investors can require that the expected return of the portfolio should not be less than a tolerable level, and in the meantime minimize the occurrence chance of the concerned low return level. That is, the investors can select the portfolio based on the following chance minimization model:

$$\left\{\begin{array}{l} \min \mathcal{M}\left\{\xi_1 x_1+\xi_2 x_2+\cdots+\xi_n x_n \leq d\right\} \\ \text{subject to:} \\ \qquad E[\xi_1 x_1+\xi_2 x_2+\cdots+\xi_n x_n] \geq a \\ \qquad x_1+x_2+\cdots+x_n=1 \\ \qquad x_i \geq 0, \quad i=1,2,\cdots,n \end{array}\right. \tag{4.46}$$

where d is the concerned low return level and a the preset minimum expected return that the investors can accept.

Let us recall the definition of risk curve. The curve

$$R(x_1, x_2, \cdots, x_n; r) = \mathcal{M}\{r_f - (\xi_1 x_1 + \xi_2 x_2 + \cdots + \xi_n x_n) \geq r\}, \forall r \geq 0$$

is called the risk curve of the portfolio, where r_f is the risk-free interest rate. If r degenerates to one specific number r_0, then the risk curve becomes

$$\begin{aligned} R(x_1, x_2, \cdots, x_n; r_0) &= \mathcal{M}\{r_f - (\xi_1 x_1 + \xi_2 x_2 + \cdots + \xi_n x_n) \geq r_0\} \\ &= \mathcal{M}\{\xi_1 x_1 + \xi_2 x_2 + \cdots + \xi_n x_n \leq r_f - r_0\} \end{aligned}$$

which is just the risk definition of occurrence chance of a sensitive low return event (or say sensitive high loss event). It is clear that $r_f - r_0 = d$.

If the investors pre-give a confidence level α, then their objective should be to minimize the maximum loss that is likely to occur at this confidence level. We call the maximum loss that is likely to occur at a preset confidence level the value-at-risk-in-uncertainty.

Definition 4.13. *Let ξ denote an uncertain return of a portfolio, and r_f the risk-free interest rate. Then Value-at-Risk-in-Uncertainty* (VaRU) *is defined as*

$$\text{VaRU}(\gamma) = \sup\{\bar{r} | \mathcal{M}\{r_f - \xi \geq \bar{r}\} \leq 1 - \gamma\}. \tag{4.47}$$

where γ is the pre-determined confidence level.

For example, if VaRU(99%) = 2%, it means that there is only a 1% occurrence chance that the portfolio return rate will drop more than 2% below the risk-free return rate. It is easy to see that VaRU is in fact an inverse version of the risk definition of the chance of a portfolio return below a specific disastrous low return level.

If the investors adopt VaRU as the investment risk, then they will ask the expected return of the portfolio not be less than a preset expected value, and in the meantime minimize the VaRU value. The chance minimization model becomes VaRU minimization model as follows:

$$\begin{cases} \min \bar{r} \\ \text{subject to:} \\ \qquad \mathcal{M}\{r_f - (\xi_1 x_1 + \xi_2 x_2 + \cdots + \xi_n x_n) \geq \bar{r}\} \leq 1 - \gamma \\ \qquad E[\xi_1 x_1 + \xi_2 x_2 + \cdots + \xi_n x_n] \geq a \\ \qquad x_1 + x_2 + \cdots + x_n = 1 \\ \qquad x_i \geq 0, \quad i = 1, 2, \cdots, n \end{cases} \tag{4.48}$$

where a is the pre-set tolerable minimum expected return, γ the pre-determined confidence level, and $\bar{r}$ the VaRU defined as

$$\sup\{\bar{r} | \mathcal{M}\{r_f - (\xi_1 x_1 + \xi_2 x_2 + \cdots + \xi_n x_n) \geq \bar{r}\} \leq 1 - \gamma\}.$$

It is seen that the VaRU minimization model (4.48) can be regarded as another version of chance minimization model (4.46).

Mathematically, Model (4.48) is a minmax model because it is equivalent to

$$\begin{cases} \min_{x_1,x_2,\cdots,x_n} \max_{\bar{r}} \bar{r} \\ \text{subject to:} \\ \qquad \mathcal{M}\{r_f - (\xi_1 x_1 + \xi_2 x_2 + \cdots + \xi_n x_n) \geq \bar{r}\} \leq 1 - \gamma \\ \qquad E[\xi_1 x_1 + \xi_2 x_2 + \cdots + \xi_n x_n] \geq a \\ \qquad x_1 + x_2 + \cdots + x_n = 1 \\ \qquad x_i \geq 0, \quad i = 1, 2, \cdots, n \end{cases} \tag{4.49}$$

where $\max \bar{r}$ is the VaRU value.

4.4.2 Crisp Equivalent

In some special cases, we can convert the chance minimization model into its crisp equivalent.

Example 4.21. When people invest, the chance of portfolio return equal to or lower than a sensitive disastrous low return level d should always be required to be less than 0.5. Thus, when security returns are regarded to be all normal uncertain variables $\xi_i \sim \mathcal{N}(e_i, \sigma_i)$, Model (4.46) becomes

$$\begin{cases} \max \left(\sum_{i=1}^{n} e_i x_i - d \right) \Big/ \sum_{i=1}^{n} \sigma_i x_i \\ \text{subject to:} \\ \qquad d \leq e_1 x_1 + e_2 x_2 + \cdots + e_n x_n \\ \qquad e_1 x_1 + e_2 x_2 + \cdots + e_n x_n \geq a \\ \qquad x_1 + x_2 + \cdots + x_n = 1 \\ \qquad x_i \geq 0, \quad i = 1, 2, \cdots, n. \end{cases} \tag{4.50}$$

Model (4.50) can easily be obtained because the sum of weighted normal uncertain variables is still a normal uncertain variable. Please note that to minimize the chance

$$\left(1 + \exp\left(\pi \left(\sum_{i=1}^{n} e_i x_i - d \right) \Big/ \sqrt{3} \sum_{i=1}^{n} \sigma_i x_i \right) \right)^{-1}$$

we just need to maximize

$$\left(\sum_{i=1}^{n} e_i x_i - d\right) \Big/ \sum_{i=1}^{n} \sigma_i x_i.$$

A constraint $d \le e_1x_1 + e_2x_2 + \cdots + e_nx_n$ is added because the chance of portfolio return equal to or less than the concerned disaster level d should be less than 0.5.

Example 4.22. When security returns are all linear uncertain variables, Model (4.46) becomes

$$\begin{cases} \min \left(d - \sum_{i=1}^{n} a_i x_i\right) \Big/ \sum_{i=1}^{n} (b_i - a_i) x_i \\ \text{subject to:} \\ \qquad \sum_{i=1}^{n} (a_i + b_i) x_i \ge 2a \\ \qquad x_1 + x_2 + \cdots + x_n = 1 \\ \qquad x_i \ge 0, \quad i = 1, 2, \cdots, n. \end{cases} \tag{4.51}$$

Theorem 4.16. *Let Φ_i denote the uncertainty distributions of the i-th security return rates $\xi_i, i = 1, 2, \cdots, n$, respectively. Then the* VaRU *minimization model can be transformed into the following linear model:*

$$\begin{cases} \min r_f - x_1\Phi_1^{-1}(1-\gamma) - x_2\Phi_2^{-1}(1-\gamma) + \cdots - x_n\Phi_n^{-1}(1-\gamma) \\ \textit{subject to:} \\ \qquad x_1E[\xi_1] + x_2E[\xi_2] + \cdots + x_nE[\xi_n] \ge a \\ \qquad x_1 + x_2 + \cdots + x_n = 1 \\ \qquad x_i \ge 0, \quad i = 1, 2, \cdots, n. \end{cases} \tag{4.52}$$

Proof: The objective function can be obtained directly from Theorem 4.6 and the monotonicity property of the uncertain measure, and the constraint can be obtained from the linearity property of expected value of the uncertain measure.

Example 4.23. Suppose the return rates of the i-th securities are normal uncertain variables $\xi_i \sim \mathcal{N}(e_i, \sigma_i), i = 1, 2, \cdots, m$, and the return rates of the j-th securities are linear uncertain variables $\xi_j = \mathcal{L}(a_j, b_j), j = m+1, m+2, \cdots, n$, respectively. Since the confidence level $\gamma > 0.5$ and $1 - \gamma < 0.5$, Model (4.48) can be transformed into the following form:

$$\begin{cases} \min r_f - \sum_{i=1}^{m} \left(e_i - \frac{\sqrt{3}\sigma_i}{\pi} \ln \frac{\gamma}{1-\gamma} \right) x_i - \\ \qquad \gamma \sum_{i=m+1}^{n} a_i x_i - (1-\gamma) \sum_{i=m+1}^{n} b_i x_i \\ \text{subject to:} \\ \qquad \sum_{i=1}^{m} e_i x_i + \sum_{i=m+1}^{n} \frac{1}{2}(b_i x_i + a_i x_i) \ge a \\ \qquad x_1 + x_2 + \cdots + x_n = 1 \\ \qquad x_i \ge 0, \quad i = 1, 2, \cdots, n. \end{cases} \tag{4.53}$$

4.4.3 An Example

Suppose an investor wants to choose an optimal portfolio from ten securities whose returns are believed to be normal uncertain variables. The prediction of the return rates of the ten securities is given in Table 4.7. The risk-free interest rate is 0.01. Suppose the minimum expected return the investor can accept is 0.03, and the investor wants to minimize potential loss at confidence level 0.95. Then according to the VaRU minimization selection idea , we build the model as follows:

$$\begin{cases} \min \bar{r} \\ \text{subject to:} \\ \qquad \mathcal{M}\{0.01 - (\xi_1 x_1 + \xi_2 x_2 + \cdots + \xi_{10} x_{10}) \ge \bar{r}\} \le 0.05 \\ \qquad E[\xi_1 x_1 + \xi_2 x_2 + \cdots + \xi_{10} x_{10}] \ge 0.03 \\ \qquad x_1 + x_2 + \cdots + x_{10} = 1 \\ \qquad x_i \ge 0, \quad i = 1, 2, \cdots, 10. \end{cases} \tag{4.54}$$

Table 4.7 Normal Uncertain Return Rates of 10 Securities

Security i	$\xi_i \sim \mathcal{N}(e_i, \sigma_i)$	Security i	$\xi_i \sim \mathcal{N}(e_i, \sigma_i)$
1	$\mathcal{N}(0.033, 0.06)$	6	$\mathcal{N}(0.035, 0.043)$
2	$\mathcal{N}(0.03, 0.065)$	7	$\mathcal{N}(0.032, 0.08)$
3	$\mathcal{N}(0.034, 0.067)$	8	$\mathcal{N}(0.036, 0.062)$
4	$\mathcal{N}(0.04, 0.08)$	9	$\mathcal{N}(0.026, 0.045)$
5	$\mathcal{N}(0.031, 0.061)$	10	$\mathcal{N}(0.028, 0.032)$

Table 4.8 Allocation of Money to Ten Securities

Security i	1	2	3	4	5
Allocation of money	0.00%	0.00 %	0.00 %	0.00%	0.00%
Security i	6	7	8	9	10
Allocation of money	28.57%	0.00 %	0.00%	0.00%	71.43%

According to Model (4.52), we change Model (4.54) into the following form:

$$\begin{cases} \min 0.01 - \sum_{i=1}^{10} \left(e_i - \frac{\sqrt{3}\sigma_i}{\pi} \ln \frac{0.95}{1-0.95} \right) x_i \\ \text{subject to:} \\ \qquad \sum_{i=1}^{10} e_i x_i \geq 0.03 \\ \qquad x_1 + x_2 + \cdots + x_{10} = 1 \\ \qquad x_i \geq 0, \quad i = 1, 2, \cdots, 10. \end{cases} \tag{4.55}$$

Using "Solver" in "Excel", we obtain the optimal portfolio shown in Table 4.8. The objective is 0.037. That is, at chance 5%, the portfolio return rate will drop at most 0.037 below the risk-free interest rate.

4.5 Mean-Variance Model

4.5.1 Mean-Variance Model

As a counter part of probabilistic mean-variance model and credibilistic mean-variance model, we provide here the mean-variance model for portfolio selection with uncertain returns.

Let ξ_i represent the uncertain returns of the i-th securities and x_i the investment proportions in the i-th securities $i = 1, 2, \cdots, n$, respectively. Suppose the uncertainty distributions of security returns are all symmetrical. When expected return is used to represent the investment return and variance as the investment risk, the optimal portfolio should be the one whose variance is not greater than the preset level and in the meantime whose expected return is the maximal; or the optimal portfolio should be the one whose expected return is not less than the preset level and in the meantime whose variance is the minimal. Let γ be the preset variance level the investors can tolerate. Then the mean-variance selection model is expressed as follows:

$$\begin{cases} \max E[x_1\xi_1 + x_2\xi_2 + \cdots + x_n\xi_n] \\ \text{subject to:} \\ \qquad V[x_1\xi_1 + x_2\xi_2 + \cdots + x_n\xi_n] \leq \gamma \\ \qquad x_1 + x_2 + \cdots + x_n = 1 \\ \qquad x_i \geq 0, \quad i = 1, 2, \cdots, n \end{cases} \tag{4.56}$$

where E denotes the expected value operator, and V the variance operator of the uncertain variables.

When the investors pre-give a minimum expected return that they can tolerate, the mean-variance model becomes:

$$\begin{cases} \min V[x_1\xi_1 + x_2\xi_2 + \cdots + x_n\xi_n] \\ \text{subject to:} \\ \qquad E[x_1\xi_1 + x_2\xi_2 + \cdots + x_n\xi_n] \geq \lambda \\ \qquad x_1 + x_2 + \cdots + x_n = 1 \\ \qquad x_i \geq 0, \quad i = 1, 2, \cdots, n \end{cases} \tag{4.57}$$

where λ represents the minimum expected return the investors feel satisfactory.

It can be seen from Models (4.56) and (4.57) that if we change the preset variance value or expected value, we will get different optimal solution. A portfolio is efficient if it is impossible to obtain higher expected return with no greater variance value, or it is impossible to obtain less variance value with no less expected return. All efficient portfolios make up the efficient frontier. An efficient portfolio is in fact a solution of the following optimization model with two objectives:

$$\begin{cases} \max E[x_1\xi_1 + x_2\xi_2 + \cdots + x_n\xi_n] \\ \min V[x_1\xi_1 + x_2\xi_2 + \cdots + x_n\xi_n] \\ \text{subject to:} \\ \qquad x_1 + x_2 + \cdots + x_n = 1 \\ \qquad x_i \geq 0, \quad i = 1, 2, \cdots, n. \end{cases} \tag{4.58}$$

Different investors will find different optimal portfolios from the efficient frontier according to their own preferences to risk aversion, i.e., tradeoff of variance and expected return.

4.5.2 Crisp Equivalent

According to the properties of normal uncertain variable and linear uncertain variable, the crisp equivalents of uncertain mean-variance model in the special cases are given when all the security returns are normal uncertain variables or when all the security returns are linear uncertain variables.

When all the security returns are normal uncertain variables $\xi_i \sim \mathcal{N}(e_i, \sigma_i)$, the uncertainty distributions of portfolio returns are symmetrical. Since the weighted sum of normal uncertain variables is still a normal uncertain variable, Model (4.56) becomes

$$\begin{cases} \max e_1x_1 + e_2x_2 + \cdots + e_nx_n \\ \text{subject to:} \\ \qquad \sigma_1x_1 + \sigma_2x_2 + \cdots + \sigma_nx_n \leq \gamma \\ \qquad x_1 + x_2 + \cdots + x_n = 1 \\ \qquad x_i \geq 0, \quad i = 1, 2, \cdots, n. \end{cases} \tag{4.59}$$

And Model (4.57) becomes

$$\begin{cases} \min \sigma_1x_1 + \sigma_2x_2 + \cdots + \sigma_nx_n \\ \text{subject to:} \\ \qquad e_1x_1 + e_2x_2 + \cdots + e_nx_n \geq \lambda \\ \qquad x_1 + x_2 + \cdots + x_n = 1 \\ \qquad x_i \geq 0, \quad i = 1, 2, \cdots, n. \end{cases} \tag{4.60}$$

When all the security returns are linear uncertain variables $\mathcal{L}(a_i, b_i)$, the uncertainty distributions of portfolio returns are symmetrical. Since the weighted sum of linear uncertain variables is still a linear uncertain variable, Model (4.56) becomes

$$\begin{cases} \max(a_1 + b_1)x_1 + (a_2 + b_2)x_2 + \cdots + (a_n + b_n)x_n \\ \text{subject to:} \\ \qquad (b_1 - a_1)x_1 + (b_2 - a_2)x_2 + \cdots + (b_n - a_n)x_n \leq \sqrt{12\gamma} \\ \qquad x_1 + x_2 + \cdots + x_n = 1 \\ \qquad x_i \geq 0, \quad i = 1, 2, \cdots, n. \end{cases} \tag{4.61}$$

And Model (4.57) becomes

$$\begin{cases} \min(b_1 - a_1)x_1 + (b_2 - a_2)x_2 + \cdots + (b_n - a_n)x_n \\ \text{subject to:} \\ \qquad (a_1 + b_1)x_1 + (a_2 + b_2)x_2 + \cdots + (a_n + b_n)x_n \geq 2\lambda \\ \qquad x_1 + x_2 + \cdots + x_n = 1 \\ \qquad x_i \geq 0, \quad i = 1, 2, \cdots, n. \end{cases} \tag{4.62}$$

4.5.3 A Solution Algorithm

When security returns are different types of uncertain variables with symmetrical uncertainty distributions, we can use 9999 Method to calculate the expected and variance values of portfolio returns and then use traditional method or integrate the calculation results into the GA to find the optimal portfolio. Calculation of expected value has been introduced in 9999 Method B. Here, we introduce the way for calculating variance via 9999 method.

Calculation of Variance

Let ξ_i represent the i-th security returns with symmetrical uncertainty distributions Φ_i, and x_i the investment proportions in securities $i, i = 1, 2, \cdots, n$, respectively. According to the definition of variance of uncertain variable and Equation (4.20), the variance of the portfolio return $\sum_{i=1}^{n} \xi_i x_i$ can be calculated via

$$V\Big[\sum_{i=1}^{n} \xi_i x_i\Big] = 2\int_e^{+\infty} (r - e)(1 - \Psi(r) + \Psi(2e - r))\mathrm{d}r$$

where Ψ is the uncertainty distribution of the uncertain portfolio return $\sum_{i=1}^{n} \xi_i x_i$, and e the expected value of the portfolio return.

Recall the 9999 Method A in page 124. Let Φ_i represent the uncertainty distributions of the i-th securities, $i = 1, 2, \cdots, n$, respectively. The uncertainty distribution Ψ of the portfolio return $\sum_{i=1}^{n} \xi_i x_i$ can be represented on a computer as follows:

α_j	0.0001	0.0002	$\cdots$	$0.0001j$	$\cdots$	0.9999
$\Phi_1^{-1}(\alpha_j)$	$t_{1/1}$	$t_{1/2}$	$\cdots$	$t_{1/j}$	$\cdots$	$t_{1/9999}$
$\Phi_2^{-1}(\alpha_j)$	$t_{2/1}$	$t_{2/2}$	$\cdots$	$t_{2/j}$	$\cdots$	$t_{2/9999}$
$\cdots$	$\cdots$	$\cdots$	$\cdots$	$\cdots$	$\cdots$	$\cdots$
$\Phi_n^{-1}(\alpha_j)$	$t_{n/1}$	$t_{n/2}$	$\cdots$	$t_{n/j}$	$\cdots$	$t_{n/9999}$
$\Psi^{-1}(\alpha_j)$	$\sum_{i=1}^{n} x_i t_{i/1}$	$\sum_{i=1}^{n} x_i t_{i/2}$	$\cdots$	$\sum_{i=1}^{n} x_i t_{i/j}$	$\cdots$	$\sum_{i=1}^{n} x_i t_{i/9999}$

(4.63)

Since the security returns are all symmetrical,

$$\sum_{i=1}^{n} x_i t_{i/j} = e \quad \text{when} \quad j = 5000. \tag{4.64}$$

Thus,

$$\begin{aligned} V\Big[\sum_{i=1}^{n} \xi_i x_i\Big] &= 2\int_e^{+\infty} (r - e)(1 - \Psi(r) + \Psi(2e - r))\mathrm{d}r \\ &= 4\sum_{j=5000}^{9999} (r_j - e)(1 - 0.0001j)(r_j - r_{j-1}) \end{aligned} \tag{4.65}$$

where $r_j = \sum_{i=1}^{n} x_i t_{i/j}$.

Genetic Algorithm

When the expected and variance values of portfolio returns have been calculated by Equations (4.64) and (4.65), respectively, we can either use traditional ways to find the optimal solution or integrate the calculation results into GA to find the optimal portfolio. Below is the summary of GA for finding the optimal solution of the mean-variance model (4.56). The optimal solution of Model (4.57) can be obtained in the similar way.

Step 1. Initialize *pop_size* chromosomes. Use Equation (4.65) to calculate the variance and check the constraint.
Step 2. Calculate the objective values for all chromosomes.
Step 3. Give the rank order of the chromosomes according to the objective values, and compute the values of the rank-based evaluation function of the chromosomes.
Step 4. Compute the fitness of each chromosome according to the rank-based-evaluation function.
Step 5. Select the chromosomes by spinning the roulette wheel.
Step 6. Update the chromosomes by crossover and mutation operations. Use Equation (4.65) to calculate the variance when checking the constraint.
Step 7. Repeat the second to the sixth steps for a given number of cycles.
Step 8. Take the best chromosome as the solution of portfolio selection.

4.5.4 An Example

Suppose an investor wants to select his/her portfolio from the ten securities whose return rates are given in Table 4.9. If the investor adopts the mean-variance selection idea, and sets the minimum expected return at 0.07. Then

Table 4.9 Uncertain Return Rates of 10 Securities

Security i	$\xi_i = \mathcal{L}(a_i, b_i)$	Security i	$\xi_i \sim \mathcal{N}(e_i, \sigma_i)$
1	$\mathcal{L}(-0.1, 0.2)$	6	$\mathcal{N}(0.06, 0.1)$
2	$\mathcal{L}(-0.12, 0.26)$	7	$\mathcal{N}(0.09, 0.14)$
3	$\mathcal{L}(-0.2, 0.38)$	8	$\mathcal{N}(0.05, 0.08)$
4	$\mathcal{L}(-0.11, 0.23)$	9	$\mathcal{N}(0.08, 0.12)$
5	$\mathcal{L}(-0.08, 0.18)$	10	$\mathcal{N}(0.1, 0.18)$

Table 4.10 Allocation of Money to Ten Securities

Security i	1	2	3	4	5
Allocation of money	0.00%	0.00 %	0.00%	0.00%	0.00%
Security i	6	7	8	9	10
Allocation of money	0.00%	33.33 %	66.67%	%	0.00%

according to the mean-variance model, the investor should select the portfolio according to the following model:

$$\begin{cases} \min V[\xi_1 x_1 + \xi_2 x_2 + \cdots + \xi_{10} x_{10}] \\ \text{subject to:} \\ \qquad E[\xi_1 x_1 + \xi_2 x_2 + \cdots + \xi_{10} x_{10}] \geq 0.07 \\ \qquad x_1 + x_2 + \cdots + x_{10} = 1 \\ \qquad x_i \geq 0, \quad i = 1, 2, \cdots, 10. \end{cases} \tag{4.66}$$

According to the properties of linear uncertain variable and normal uncertain variable, we change Model (4.66) into the following crisp form:

$$\begin{cases} \min V[\xi_1 x_1 + \xi_2 x_2 + \cdots + \xi_{10} x_{10}] \\ \text{subject to:} \\ \qquad 0.05x_1 + 0.07x_2 + 0.09x_3 + 0.06x_4 + 0.05x_5 + \\ \qquad 0.06x_6 + 0.09x_7 + 0.05x_8 + 0.08x_9 + 0.1x_{10} \geq 0.07 \\ \qquad x_1 + x_2 + \cdots + x_{10} = 1 \\ \qquad x_i \geq 0, \quad i = 1, 2, \cdots, 10. \end{cases} \tag{4.67}$$

We use 999999 Method to calculate the variance values of portfolios and integrate the calculation results into the GA. A run of the GA ($P_m = 0.2, P_c = 0.3, pop\_size = 30, a = 0.05$) with 10000 generations shows that to minimize variance at the constraint that the expected return should not be lower than 0.07, the investor should allocate his/her money according to Table 4.10. The minimum variance value is 0.0114.

Chapter 5
Model Varieties

There is a widely accepted rule that diversification is an efficient way for reducing likely loss of an investment. The idea is reflected in a famous saying that "one should not put all the eggs into one basket". In portfolio selection, this means the investors should not allocate all their money to just a few securities. Placing all the money in only a few companies could lead to financial disaster. One example of this is the nearly 80 percent decline in the Nasdaq market index with many high-tech companies from March 2000 to October 2002.

Diversification works because it is rare that all the securities will perform poorly simultaneously, nor is it common that all the securities will perform same poorly at the same time. In fact, there is an unstated worry hidden in the diversification rule. This worry is that people have to make decision based on their prediction of security returns, but their prediction may be wrong sometimes, or some contingency may change the security returns. If that is the case, however, it is rare that prediction about all security returns is wrong. Thus, a diversified investment would lose less than a concentrative investment.

In the area of probabilistic portfolio selection, it has been found that optimal mean-variance portfolio is often extremely concentrated on a few securities [43]. Thus, Kapur and Kesavan [44], Kapur [45], Fang, Rajasekera and Tsao [17], Jana, Roy and Mazumder [43] suggested the diversified mean-variance models. Similarly, the optimal portfolio produced by other types of selection criteria such as the mean-risk model, β-return-risk model and probability minimization model may also be quite concentrative on a few securities. The similar concentrative optimal portfolio is also found in credibilitic portfolio selection and uncertain portfolio selection. Thus, for conservative investors, diversification versions of portfolio models for various selection criteria are needed.

We must point out that diversification proposed in this chapter does not have the same sense as the diversification implied in the former chapters. In the former chapters, securities are diversified in the sense of lowering the risk

X. Huang: Portfolio Analysis: From Probab. to Credibilistic, STUDFUZZ 250, pp. 157–172.
springerlink.com

curve, reducing the chance value of a concerned low return event, or reducing the variance value of a portfolio. Diversification in this chapter is just a compromise solution to people's dilemma that they have to make decision based on their prediction of security returns and in the meantime they dare not totally believe their prediction.

5.1 Entropy and Diversification

Shannon entropy was usually applied to measure the uncertain degree of a portfolio return, see [78, 79, 72]. But it can also be used to measure the diversification degree of an investment. To understand it, let us recall the definition of the entropy first.

Let η be a discrete random variable taking values a_i at probabilities $p_i, i = 1, 2, \cdots, n$, respectively. Then its entropy is defined by [85]

$$H[\eta] = -\sum_{i=1}^{n} p_i \ln p_i.$$

The entropy measures how equal the probabilities $p_1, p_2, \cdots, p_n$ are among themselves. The greater value the entropy, the closer the random variable is to the equi-probable random variable, or the vice versa. The entropy value will reach its minimum of 0 if and only if there exists an index k such that $p_k = 1$, and will reach its maximum of $\ln n$ if and only if $p_i \equiv 1/n$ for all $i = 1, 2, \cdots, n$.

Notice that in portfolio selection problem, the uniform degree of the investment proportions is also a token of diversification degree of the investment. The more uniform the investment proportions in the securities are, the more dispersed the investment is. Furthermore, the investment proportions in securities i, denoted by x_i, meet the same requirement for probabilities p_i, i.e., $x_i \geq 0$ and $\sum_{i=1}^{n} x_i = 1, i = 1, 2, \cdots, n$. Thus, we replace the probabilities in Shannon entropy by investment proportions and make use of the properties of entropy to reflect the diversification degree of a portfolio. For clarity, this entropy is called the proportion entropy.

Definition 5.1 *(Pan, Huang [73]) Let x_i denote the investment proportions in the i-th securities, $i = 1, 2, \cdots, n$, respectively. Then the proportion entropy is*

$$H = -\sum_{i=1}^{n} x_i \ln x_i. \tag{5.1}$$

For example, for the portfolios containing three alternative securities A, B, and C, the proportion entropy values are given in Table 5.1. For portfolio 1, its proportion entropy is 0 and the investment proportion of security C reaches the maximum value $x_3 = 1$, which implies that only security C is allocated the money and the other two securities are allocated nothing. In this case the investment is extremely concentrative. For portfolio 2, its proportion entropy becomes bigger than the proportion entropy of portfolio 1 and the investment proportions of securities B and C are greater than zero, which implies that two securities B and C are allocated some money. The investment in portfolio 2 is more dispersed than in portfolio 1. For portfolios 3 to 6, the proportion entropy values become bigger and bigger and the investment proportions of all the three securities possess positive numbers, which implies that all the securities are allocated some money. The investments are more dispersed than portfolios 1 or 2. The portfolio becomes dispersed when the proportion entropy value becomes big.

Table 5.1 Proportion Entropy and Diversification

Portfolio	Proportion in A (x_1)	Proportion in B (x_2)	Proportion in C (x_3)	Proportion Entropy
1	0.0	0.0	1.0	0.000
2	0.0	0.1	0.9	0.325
3	0.1	0.1	0.8	0.639
4	0.1	0.2	0.7	0.802
5	0.2	0.3	0.5	1.030
6	0.33	0.33	0.34	1.099

From the properties of shannon entropy, it is easy to get that

(1) When the proportion entropy value is 0, the portfolio will be extremely concentrative. The portfolio contains only one security.

(2) When the proportion entropy value reaches its maximum value of $\ln n$, the portfolio will be most dispersed. The portfolio contains all the securities, and the money is allocated evenly to all the securities.

(3) The greater value the proportion entropy takes, the more dispersed the portfolio is. Thus, the investors can choose the entropy value from the interval $(0, \ln n)$ according to their own requirement for diversification.

Note that the proportion entropy itself does not use any information whatever about the uncertainty of the security returns. Therefore, the proportion entropy is not a risk measure all by itself. It has to be combined with a chosen measure of risk such that it can serve as a complementary means to reduce risk. In addition, it is not suggested that the more dispersed the portfolio, the better the portfolio. The investors should choose the entropy value from the interval $(0, \ln n)$ according to their own requirement for diversification.

5.2 Mean-Risk Diversification Models

Probabilistic Mean-Risk Diversification Model

If the investors adopt risk curve as the risk measure and accept mean-risk selection criterion and preset a diversification degree β, then in the situation where security returns are random variables, the mean-risk diversification model is as follows:

$$\begin{cases} \max E[\xi_1 x_1 + \xi_2 x_2 + \cdots + \xi_n x_n] \\ \text{subject to:} \\ \qquad R(x_1, x_2, \cdots, x_n; r) \le \alpha(r), \ \forall r \ge 0 \\ \qquad -x_1 \ln x_1 - x_2 \ln x_2 - \cdots - x_n \ln x_n \ge \beta \\ \qquad x_1 + x_2 + \cdots + x_n = 1 \\ \qquad x_i \ge 0, \quad i = 1, 2, \cdots, n, \end{cases} \tag{5.2}$$

where ξ_i denote the random returns of the i-th securities, x_i the investment proportions in the i-th securities, E the expected value operator of random variables, $\alpha(r)$ the investors' confidence curve, $R(x_1, x_2, \cdots, x_n; r)$ the risk curve of the random portfolio $(x_1, x_2, \cdots, x_n)$ defined as

$$R(x_1, x_2, \cdots, x_n; r) = \Pr\{r_f - (\xi_1 x_1 + \xi_2 x_2 + \cdots + \xi_n x_n) \ge r\}$$

in which r_f is the risk-free interest rate. In the mean-risk diversification model (5.2), the investors do not only require that the risk curve should be below the confidence curve but also require that the portfolio be dispersed enough. The investors now are more conservative than the investors who adopt the probabilistic mean-risk model (2.16).

Credibilistic Mean-Risk Diversification Model

In the case when security returns are fuzzy numbers, the mean-risk diversification model is as follows:

$$\begin{cases} \max E[\xi_1 x_1 + \xi_2 x_2 + \cdots + \xi_n x_n] \\ \text{subject to:} \\ \qquad R(x_1, x_2, \cdots, x_n; r) \le \alpha(r), \ \forall r \ge 0 \\ \qquad -x_1 \ln x_1 - x_2 \ln x_2 - \cdots - x_n \ln x_n \ge \beta \\ \qquad x_1 + x_2 + \cdots + x_n = 1 \\ \qquad x_i \ge 0, \quad i = 1, 2, \cdots, n, \end{cases} \tag{5.3}$$

where ξ_i denote the fuzzy returns of the i-th securities, E the expected value operator of the fuzzy variables, $R(x_1, x_2, \cdots, x_n; r)$ the risk curve of the fuzzy portfolio $(x_1, x_2, \cdots, x_n)$ defined as

$$R(x_1, x_2, \cdots, x_n; r) = \mathrm{Cr}\{r_f - (\xi_1 x_1 + \xi_2 x_2 + \cdots + \xi_n x_n) \geq r\}$$

in which r_f is the risk-free interest rate.

Uncertain Mean-Risk Diversification Model

In the case when security returns are uncertain variables, the mean-risk diversification model is:

$$\begin{cases} \max E[\xi_1 x_1 + \xi_2 x_2 + \cdots + \xi_n x_n] \\ \text{subject to:} \\ \qquad R(x_1, x_2, \cdots, x_n; r) \leq \alpha(r), \ \forall r \geq 0 \\ \qquad -x_1 \ln x_1 - x_2 \ln x_2 - \cdots - x_n \ln x_n \geq \beta \\ \qquad x_1 + x_2 + \cdots + x_n = 1 \\ \qquad x_i \geq 0, \quad i = 1, 2, \cdots, n, \end{cases} \tag{5.4}$$

where ξ_i denote the uncertain returns of the i-th securities, E the expected value operator of the uncertain variables, $R(x_1, x_2, \cdots, x_n; r)$ the risk curve of the uncertain portfolio $(x_1, x_2, \cdots, x_n)$ defined as

$$R(x_1, x_2, \cdots, x_n; r) = \mathcal{M}\{r_f - (\xi_1 x_1 + \xi_2 x_2 + \cdots + \xi_n x_n) \geq r\}$$

in which r_f is the risk-free interest rate.

Application Examples

Example of Probabilistic Mean-Risk Diversification Model

see also Section 2.2.4

Recall the former mentioned six alternative securities Hundsun (600570), Tianjin (600821), Wanwei (600063), Sany (600031), Baosteel (600019), and Tianchuang (600791) whose monthly returns are given in Section 2.2.4 in Table 2.6. Suppose the investors would like to select the portfolio from them and they adopt mean-risk diversification selection criterion. The monthly risk-free interest rate is still $r_f = 0.003$, and the investors' confidence curve is the same as follows:

$$\alpha(r) = \begin{cases} -1.25r + 0.25, & \text{when} \quad 0 \leq r \leq 0.12 \\ -0.5r + 0.16, & \text{when} \quad 0.12 \leq r \leq 0.3 \\ 0.01, & \text{when} \quad r \geq 0.3. \end{cases}$$

If the investors now ask that the proportion entropy value should not be less than 1.2, then their mean-risk diversification selection model is as follows:

$$\begin{cases} \max E[\xi_1 x_1 + \xi_2 x_2 + \xi_3 x_3 + \xi_4 x_4 + \xi_5 x_5 + \xi_6 x_6] \\ \text{subject to:} \\ \qquad R(x_1, x_2, \cdots, x_6; r) \le \alpha(r), \ \forall r \ge 0 \\ \qquad -x_1 \ln x_1 - x_2 \ln x_2 - \cdots - x_6 \ln x_6 \ge 1.2 \\ \qquad x_1 + x_2 + x_3 + x_4 + x_5 + x_6 = 1 \\ \qquad x_1, x_2, x_3, x_4, x_5, x_6 \ge 0 \end{cases} \tag{5.5}$$

where $\xi_1, \xi_2, \xi_3, \xi_4, \xi_5, \xi_6$ represent the random monthly returns of securities Hundsun (600570), Tianjin (600821), Wanwei(600063), Sany(600031), Baosteel(600019), and Tianchuang(600791), respectively, and

$$R(x_1, x_2, \cdots, x_6; r) = \Pr\{0.003 - (\xi_1 x_1 + \xi_2 x_2 + \cdots + \xi_6 x_6) \ge r\}.$$

A run of "Solver" in the menu "Tool" of Microsoft Excel shows that when the diversification constraint is added, the investors should assign their money according to Table 5.2. The expected return now is 0.1074. As shown in Fig. 5.1, the risk curve $R(r)$ of the dispersed model is totally below the investors' confidence curve $\alpha(r)$.

Table 5.2 Allocation of Money to Six Securities

600570	600821	600063	600031	600019	600791
20.68%	2.87%	26.69%	48.39%	0.65%	0.72%

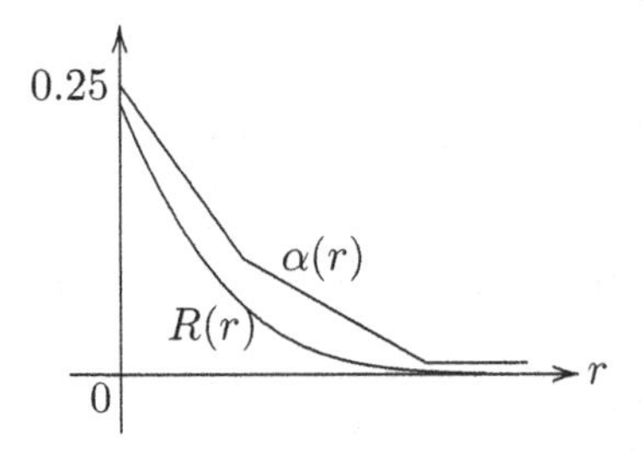

Fig. 5.1 Risk curve $R(r)$ and confidence curve $\alpha(r)$ of Model (5.5).

Compared with the optimal portfolio in Table 2.7 produced by probabilistic mean-risk model (2.17) in Subsection 2.2.4, the optimal portfolio now is much more dispersed. In Table 2.7 (also see the first row in Table 5.3), the portfolio includes only three securities without entropy constraint, while in

Table 5.2, the portfolio includes all the six securities. In the mean-risk diversification model, the risk curve and the proportion entropy are tied up to measure the risk of the portfolio. Only the portfolios whose risk curves are below the confidence curve and in the meantime whose diversification degrees are equal to or greater than the preset level are regarded to be safe. By comparison, we see that the expected value 0.1074 of the dispersed portfolio is a little lower than the expected value 0.1099 produced from the mean-risk model, which reflects the relationship between risk and return, i.e., the lower the risk, the lower the return; the higher the risk, the higher the return.

Table 5.3 Portfolios within Different Diversification Constraints

600570	600821	600063	600031	600019	600791	entr.	obj.
5.47%	0.00%	21.03%	73.5%	0.00%	0.00%	0.0	0.1099
9.1%	0.07%	17.06%	73.76%	0.00%	0.00%	0.75	0.1098
16.18%	0.81%	23.87%	58.96%	0.09%	0.1%	1.0	0.1087
20.68%	2.87%	26.69%	48.39%	0.65%	0.72%	1.2	0.1074
22.76%	7.92%	26.07%	35.84%	3.59%	3.81%	1.5	0.1039

For further comparison, we did more experiments with different proportion entropy values in the constraint. The results are given in Table 5.3. When the proportion entropy becomes bigger, the selected portfolio becomes more diversified, but in the meantime, the expected return becomes smaller.

Example of Credibilistic Mean-Risk Diversification Model

see also Section 3.2.5

Recall the application example of credibilistic mean-risk model for fuzzy portfolio selection in Subsection 3.2.5. Suppose the investors would like to select the portfolio from the securities given in Table 3.1 in Subsection 3.2.5. This time they adopt mean-risk diversification selection criterion. The monthly risk-free interest rate is still $r_f = 0.01$, and the investors' confidence curve is the same as follows:

$$\alpha(r) = \begin{cases} -2.75r + 0.43, & 0 \le r \le 0.12, \\ -0.5r + 0.16, & 0.12 \le r \le 0.3, \\ 0.01, & r \ge 0.3. \end{cases}$$

If the investors now ask that the proportion entropy value should not be less than 1.3, then their mean-risk diversification selection model is as follows:

$$
\left\{
\begin{array}{l}
\max E[\xi_1 x_1 + \xi_2 x_2 + \xi_3 x_3 + \xi_4 x_4 + \xi_5 x_5 + \xi_{10} x_{10}] \\
\text{subject to:} \\
\qquad R(x_1, x_2, \cdots, x_{10}; r) \le \alpha(r), \ \forall r \ge 0 \\
\qquad -x_1 \ln x_1 - x_2 \ln x_2 - \cdots - x_n \ln x_{10} \ge 1.3 \\
\qquad x_1 + x_2 + x_3 + x_4 + x_5 + x_{10} = 1 \\
\qquad x_1, x_2, x_3, x_4, x_5, x_{10} \ge 0
\end{array}
\right.
\tag{5.6}
$$

where

$$
R(x_1, x_2, \cdots, x_{10}; r) = \operatorname{Cr}\{0.01 - (\xi_1 x_1 + \xi_2 x_2 + \cdots + \xi_{10} x_{10}) \ge r\}.
$$

A run of "Solver" in the menu "Tool" of Microsoft Excel shows that when the diversification constraint is added, the investors should assign their money according to Table 5.4. The expected return now is 0.0417. As shown in Fig. 5.4, the risk curve $R(r)$ is totally below the confidence curve $\alpha(r)$.

Table 5.4 Allocation of Money to Ten Securities

Security i	1	2	3	4	5
Allocation of money	0.16%	0.89 %	24.94%	0.09%	0.00%
Security i	6	7	8	9	10
Allocation of money	0.39%	46.11 %	3.91 %	1.25%	22.26%

Compared with the optimal portfolio in Table 3.2 produced by credibilistic mean-risk model (3.39), the optimal portfolio now is much more dispersed. In Table 3.2 (also see the first row in Table 5.5), the portfolio includes only two securities without entropy constraint, while in Table 5.4, the portfolio includes eight securities. We also find that the expected value 0.0417 of the dispersed portfolio is a little lower than the expected value 0.0421 of the

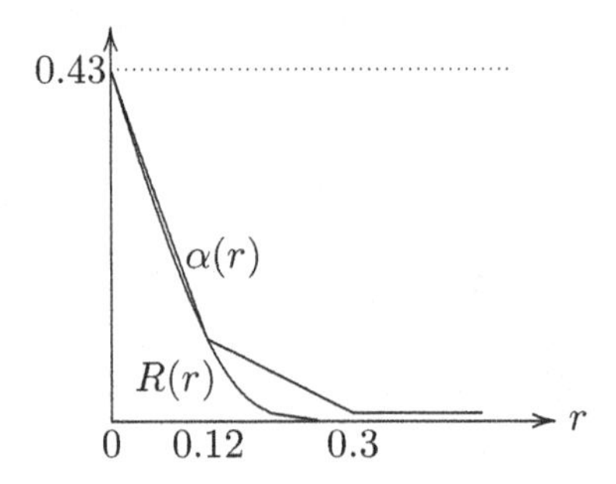

Fig. 5.2 Risk curve $R(r)$ and confidence curve $\alpha(r)$ of Model (5.6).

portfolio produced by the mean-risk model, which reflects the relationship between risk and return, i.e., the lower the risk, the lower the return; the higher the risk, the higher the return.

For further comparison, we did more experiments with different proportion entropy values in the constraint. The results are given in Table 5.5. When the proportion entropy becomes bigger, the selected portfolio becomes more diversified, but in the meantime, the expected return becomes smaller.

Table 5.5 Portfolios within Different Diversification Constraints

Security	Proportion	Proportion	Proportion	Proportion
1	0.00%	0.00%	0.16%	2.99%
2	0.00%	0.00%	0.89%	6.87%
3	0.00%	8.57%	24.94%	18.14%
4	0.00%	0.00%	0.09%	3.76%
5	0.00%	0.00%	0.00%	1.27%
6	0.00%	0.00%	0.39%	6.87%
7	78.57%	69.16%	46.11%	29.73%
8	0.00%	0.00%	3.91%	12.66%
9	0.00%	0.00%	1.25%	5.52%
10	21.43%	22.27%	22.26%	12.19%
Entropy	0	0.8	1.3	2.0
Objective	4.21%	4.21%	4.17%	3.87%

Example of Uncertain Mean-Risk Diversification Model

see also Section 4.2.5

Recall the application example of uncertain mean-risk model in Subsection 4.2.5. Suppose the investors would like to select the portfolio from the ten securities given in Table 4.3 in Subsection 4.2.5. This time they adopt mean-risk diversification selection criterion. The monthly risk-free interest rate is still $r_f = 0.01$, and the investors' confidence curve is the same as follows:

$$\alpha(r) = \begin{cases} -2.75r + 0.43, & 0 \le r \le 0.12, \\ -0.5r + 0.16, & 0.12 \le r \le 0.3, \\ 0.01, & r \ge 0.3. \end{cases}$$

If the investors now ask that the proportion entropy value should not be less than 0.8, then their mean-risk diversification selection model is as follows:

$$\begin{cases} \max E[\xi_1 x_1 + \xi_2 x_2 + \xi_3 x_3 + \xi_4 x_4 + \xi_5 x_5 + \xi_{10} x_{10}] \\ \text{subject to:} \\ \qquad R(x_1, x_2, \cdots, x_{10}; r) \le \alpha(r),\ \forall r \ge 0 \\ \qquad -x_1 \ln x_1 - x_2 \ln x_2 - \cdots - x_n \ln x_{10} \ge 0.8 \\ \qquad x_1 + x_2 + x_3 + x_4 + x_5 + x_{10} = 1 \\ \qquad x_1, x_2, x_3, x_4, x_5, x_{10} \ge 0 \end{cases} \tag{5.7}$$

where

$$R(x_1, x_2, \cdots, x_{10}; r) = \mathcal{M}\{0.01 - (\xi_1 x_1 + \xi_2 x_2 + \cdots + \xi_{10} x_{10}) \ge r\}.$$

A run of "Solver" in the menu "Tool" of Microsoft Excel shows that when the diversification constraint is added, the investors should assign their money according to Table 5.6. The expected return now is 0.041. As shown in Fig. 5.6, the risk curve $R(r)$ is totally below the confidence curve $\alpha(r)$.

Table 5.6 Allocation of Money to Ten Securities

Security i	1	2	3	4	5
Allocation of money	0.00%	0.00 %	0.00%	0.00%	0.00%
Security i	6	7	8	9	10
Allocation of money	35.56%	3.91 %	0.02 %	0.00%	60.51%

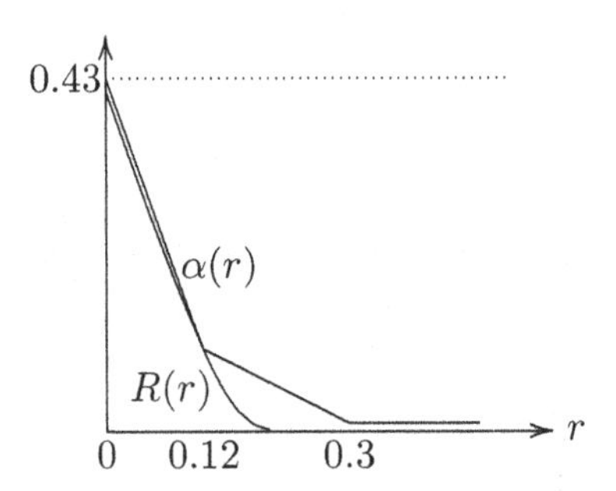

Fig. 5.3 Risk curve $R(r)$ and confidence curve $\alpha(r)$ of Model (5.7).

Compared with the optimal portfolio in Table 4.4 produced by uncertain mean-risk model (4.37), the optimal portfolio now is much more dispersed. In Table 4.4 (also see the first row in Table 5.7), the portfolio includes only two securities without entropy constraint, while in Table 5.6, the portfolio includes four securities. We also find that the expected value 0.04068 of the dispersed portfolio is a little lower than the expected value 0.0407 of the portfolio produced from the mean-risk model, which reflects the relationship between risk and return, i.e., the lower the risk, the lower the return; the higher the risk, the higher the return.

For further comparison, we did more experiments with different proportion entropy values in the constraint. The results are given in Table 5.7. When the proportion entropy becomes bigger, the selected portfolio becomes more diversified, but in the meantime, the expected return becomes smaller.

Table 5.7 Portfolios within Different Diversification Constraints

Security	Proportion	Proportion	Proportion	Proportion
1	0.00%	0.00%	0.12%	1.94%
2	0.00%	0.00%	0.43%	3.30%
3	0.00%	0.00%	0.37%	3.39%
4	0.00%	0.00%	0.55%	3.62%
5	0.00%	0.00%	1.39%	5.00%
6	38.72%	35.56%	14.63%	13.69%
7	0.00%	3.91%	15.69%	14.79%
8	0.00%	0.02%	11.58%	13.00%
9	0.00%	0.00%	7.02%	11.16%
10	61.28%	60.51%	48.21%	30.11%
Entropy	0	0.8	1.2	2.0
Objective	4.07%	4.07%	3.94%	3.62%

5.3 β-Return-Risk Diversification Models

Probabilistic β-Return-Risk Diversification Model

If the investors adopt risk curve as the risk measure and accept β-return-risk selection criterion and preset a diversification degree β, then in the situation where security returns are random variables, the β-return-risk diversification model is as follows:

$$\begin{cases} \max \bar{f} \\ \text{subject to:} \\ \qquad \Pr\{\xi_1 x_1 + \xi_2 x_2 + \cdots + \xi_n x_n \geq \bar{f}\} \\ \qquad R(x_1, x_2, \cdots, x_n; r) \leq \alpha(r), \ \forall r \geq 0 \\ \qquad -x_1 \ln x_1 - x_2 \ln x_2 - \cdots - x_n \ln x_n \geq \beta \\ \qquad x_1 + x_2 + \cdots + x_n = 1 \\ \qquad x_i \geq 0, \quad i = 1, 2, \cdots, n, \end{cases} \tag{5.8}$$

where ξ_i denote the random returns of the i-th securities, x_i the investment proportions in the i-th securities, $\alpha(r)$ the investors' confidence curve, $R(x_1, x_2, \cdots, x_n; r)$ the risk curve of the random portfolio $(x_1, x_2, \cdots, x_n)$ defined as

$$\Pr\{r_f - (\xi_1 x_1 + \xi_2 x_2 + \cdots + \xi_n x_n) \geq r\},$$

and $\bar{f}$ the β-return defined as

$$\max\{\bar{f} \mid \Pr\{\xi_1 x_1 + \xi_2 x_2 + \cdots + \xi_n x_n \geq \bar{f}\} \geq \beta\}.$$

In the β-return-risk diversification model (5.8), the investors do not only require that the risk curve should be totally below the confidence curve but also require that the portfolio be dispersed enough. The investors now are more conservative than the investors who adopt the β-return-risk model (2.19).

Credibilistic β-Return-Risk Diversification Model

If the investors adopt risk curve as the risk measure and accept β-return-risk selection criterion and preset a diversification degree β, then in the situation where security returns are fuzzy numbers, the β-return-risk diversification model is as follows:

$$\begin{cases} \max \bar{f} \\ \text{subject to:} \\ \qquad \mathrm{Cr}\{\xi_1 x_1 + \xi_2 x_2 + \cdots + \xi_n x_n \geq \bar{f}\} \\ \qquad R(x_1, x_2, \cdots, x_n; r) \leq \alpha(r), \ \forall r \geq 0 \\ \qquad -x_1 \ln x_1 - x_2 \ln x_2 - \cdots - x_n \ln x_n \geq \beta \\ \qquad x_1 + x_2 + \cdots + x_n = 1 \\ \qquad x_i \geq 0, \quad i = 1, 2, \cdots, n, \end{cases} \tag{5.9}$$

where ξ_i denote the fuzzy returns of the i-th securities, $R(x_1, x_2, \cdots, x_n; r)$ the risk curve of the fuzzy portfolio $(x_1, x_2, \cdots, x_n)$ defined as

$$\mathrm{Cr}\{r_f - (\xi_1 x_1 + \xi_2 x_2 + \cdots + \xi_n x_n) \geq r\},$$

and $\bar{f}$ the β-return defined as

$$\max\{\bar{f} \mid \mathrm{Cr}\{\xi_1 x_1 + \xi_2 x_2 + \cdots + \xi_n x_n \geq \bar{f}\} \geq \beta\}.$$

In the β-return-risk diversification model (5.9), the investors do not only require that the risk curve should be below the confidence curve but also require that the portfolio be dispersed enough. The investors now are more conservative than the investors who adopt the β-return-risk model (3.43).

Uncertain β-Return-Risk Diversification Model

In the case when security returns are uncertain variables, the β-return-risk diversification model is:

$$\begin{cases} \max \bar{f} \\ \text{subject to:} \\ \qquad \mathcal{M}\{\xi_1 x_1 + \xi_2 x_2 + \cdots + \xi_n x_n \geq \bar{f}\} \\ \qquad R(x_1, x_2, \cdots, x_n; r) \leq \alpha(r), \ \forall r \geq 0 \\ \qquad -x_1 \ln x_1 - x_2 \ln x_2 - \cdots - x_n \ln x_n \geq \beta \\ \qquad x_1 + x_2 + \cdots + x_n = 1 \\ \qquad x_i \geq 0, \quad i = 1, 2, \cdots, n, \end{cases} \tag{5.10}$$

where ξ_i denote the uncertain returns of the i-th securities, $R(x_1, x_2, \cdots, x_n; r)$ the risk curve of the uncertain portfolio $(x_1, x_2, \cdots, x_n)$ defined as

$$\mathcal{M}\{r_r - (\xi_1 x_1 + \xi_2 x_2 + \cdots + \xi_n x_n) \geq r\},$$

and $\bar{f}$ the β-return defined as

$$\max\{\bar{f} \mid \mathcal{M}\{\xi_1 x_1 + \xi_2 x_2 + \cdots + \xi_n x_n \geq \bar{f}\} \geq \beta\}.$$

5.4 Chance Minimization Diversification Models

Probability Minimization Diversification Mode

When security returns are regarded to be random variables, if the investors regard the occurrence probability of a preset loss level as the risk and adopt probability minimization selection criterion, the probability minimization diversification model is as follows:

$$\begin{cases} \min \Pr\{\xi_1 x_1 + \xi_2 x_2 + \cdots + \xi_n x_n \leq d\} \\ \text{subject to:} \\ \qquad -x_1 \ln x_1 - x_2 \ln x_2 - \cdots - x_n \ln x_n \geq \beta \\ \qquad x_1 + x_2 + \cdots + x_n = 1 \\ \qquad x_i \geq 0, \quad i = 1, 2, \cdots, n \end{cases} \tag{5.11}$$

where ξ_i denote the random returns of the i-th securities, x_i the investment proportions in the i-th securities, d the concerned disastrous return level and β the preset proportion entropy value.

If the investors take investment return into account, the probability minimization diversification model can be expressed as follows:

$$\begin{cases} \min \Pr\{\xi_1 x_1 + \xi_2 x_2 + \cdots + \xi_n x_n \le d\} \\ \text{subject to:} \\ \qquad E[\xi_1 x_1 + \xi_2 x_2 + \cdots + \xi_n x_n] \ge a \\ \qquad -x_1 \ln x_1 - x_2 \ln x_2 - \cdots - x_n \ln x_n \ge \beta \\ \qquad x_1 + x_2 + \cdots + x_n = 1 \\ \qquad x_i \ge 0, \quad i = 1, 2, \cdots, n \end{cases} \tag{5.12}$$

where E is the expected value operator of random variables and a the minimum expected return the investors can accept.

Credibility Minimization Diversification Mode

In fuzzy portfolio selection, when the investors regard the occurrence credibility of a preset loss level as the risk and adopt credibility minimization selection criterion, the credibility minimization diversification model is as follows:

$$\begin{cases} \min \mathrm{Cr}\{\xi_1 x_1 + \xi_2 x_2 + \cdots + \xi_n x_n \le d\} \\ \text{subject to:} \\ \qquad E[\xi_1 x_1 + \xi_2 x_2 + \cdots + \xi_n x_n] \ge a \\ \qquad -x_1 \ln x_1 - x_2 \ln x_2 - \cdots - x_n \ln x_n \ge \beta \\ \qquad x_1 + x_2 + \cdots + x_n = 1 \\ \qquad x_i \ge 0, \quad i = 1, 2, \cdots, n \end{cases} \tag{5.13}$$

where where ξ_i denote the fuzzy returns of the i-th securities, and E the expected value operator of the fuzzy variables.

Chance Minimization Diversification Mode

When security returns are regarded to be uncertain variables, if the investors regard the occurrence chance of a preset loss level as the risk and adopt chance minimization selection criterion, the chance minimization diversification model is as follows:

$$\begin{cases} \min \mathcal{M}\{\xi_1 x_1 + \xi_2 x_2 + \cdots + \xi_n x_n \le d\} \\ \text{subject to:} \\ \qquad E[\xi_1 x_1 + \xi_2 x_2 + \cdots + \xi_n x_n] \ge a \\ \qquad -x_1 \ln x_1 - x_2 \ln x_2 - \cdots - x_n \ln x_n \ge \beta \\ \qquad x_1 + x_2 + \cdots + x_n = 1 \\ \qquad x_i \ge 0, \quad i = 1, 2, \cdots, n \end{cases} \tag{5.14}$$

where ξ_i denote the uncertain returns of the i-th securities, and E the expected value operator of the uncertain variables.

5.5 Mean-Variance Diversification Models

Probabilistic Mean-Variance Diversification Model

If the investors adopt variance as risk measure and accept mean-variance selection criterion, but add diversification requirement to the investment, when security returns are regarded to be random variables, the selection model becomes the mean-variance diversification model as follows:

$$\begin{cases} \min V[x_1\xi_1 + x_2\xi_2 + \cdots + x_n\xi_n] \\ \text{subject to:} \\ \qquad E[x_1\xi_1 + x_2\xi_2 + \cdots + x_n\xi_n] \ge \alpha \\ \qquad -x_1 \ln x_1 - x_2 \ln x_2 - \cdots - x_n \ln x_n \ge \beta \\ \qquad x_1 + x_2 + \cdots + x_n = 1 \\ \qquad x_i \ge 0, \quad i = 1, 2, \cdots, n \end{cases} \tag{5.15}$$

where ξ_i denote the random returns of the i-th securities, x_i the investment proportions in the i-th securities, E and V the expected value operator and variance of random variables, respectively, α the minimum expected return the investor can accept and β the preset minimum entropy value.

Credibilistic Mean-Variance Diversification Model

When security returns are regarded to be fuzzy numbers, the crdibilistic mean-variance diversification model is as follows:

$$\begin{cases} \min V[x_1\xi_1 + x_2\xi_2 + \cdots + x_n\xi_n] \\ \text{subject to:} \\ \qquad E[x_1\xi_1 + x_2\xi_2 + \cdots + x_n\xi_n] \ge \alpha \\ \qquad -x_1 \ln x_1 - x_2 \ln x_2 - \cdots - x_n \ln x_n \ge \beta \\ \qquad x_1 + x_2 + \cdots + x_n = 1 \\ \qquad x_i \ge 0, \quad i = 1, 2, \cdots, n \end{cases} \tag{5.16}$$

where ξ_i denote the fuzzy returns of the i-th securities, and E and V the expected value operator and variance of fuzzy variables, respectively.

Uncertain Mean-Variance Diversification Model

When security returns are regarded to be uncertain numbers, the uncertain mean-variance diversification model is as follows:

$$\begin{cases} \min V[x_1\xi_1 + x_2\xi_2 + \cdots + x_n\xi_n] \\ \text{subject to:} \\ \qquad E[x_1\xi_1 + x_2\xi_2 + \cdots + x_n\xi_n] \geq \alpha \\ \qquad -x_1 \ln x_1 - x_2 \ln x_2 - \cdots - x_n \ln x_n \geq \beta \\ \qquad x_1 + x_2 + \cdots + x_n = 1 \\ \qquad x_i \geq 0, \quad i = 1, 2, \cdots, n \end{cases} \tag{5.17}$$

where ξ_i denote the uncertain returns of the i-th securities, and E and V the expected value operator and variance of uncertain variables, respectively.

References

[1] Abdelaziz, F.B., Aouni, B., Fayedh, R.E.: Multi-objective stochastic programming for portfolio selection. European Journal of Operational Research 177, 1811–1823 (2007)

[2] Ang, J.S.: A note on the E, SL portfolio selection model. Journal of Financial and Quantitative Analysis 10, 849–857 (1975)

[3] Arditti, F.D.: Another look at mutual fund performance. Journal of Financial and Quantitative Analysis 6, 909–912 (1971)

[4] Best, M.J., Hlouskova, J.: The efficient frontier for bounded assets. Mathematical Methods of Operations Research 52, 195–212 (2000)

[5] Bilbao-Terol, A., Pérez-Gladish, B., Arenas-Parra, M., Rodríguez-Uría, M.V.: Fuzzy compromise programming for portfolio selection. Applied Mathematics and Computation 173, 251–264 (2006)

[6] Bratley, P., Fox, B.L., Schrage, L.E.: A Guide to Simulation. Springer, New York (1987)

[7] Carlsson, C., Fullér, R., Majlender, P.: A possibilistic approach to selecting portfolios with highest utility score. Fuzzy Sets and Systems 131, 13–21 (2002)

[8] Chen, J.E., Otto, K.N.: Constructing membership functions using interpolation and measurement theory. Fuzzy Sets and Systems 73, 313–327 (1995)

[9] Choobineh, F., Branting, D.: A simple approximation for semivariance. European Journal of Operational Research 27, 364–370 (1986)

[10] Chopra, V.K., Ziembia, W.T.: The effect of errors in means, variances, and covariances on optimal portfolio choices. In: Ziembia, W.T., Mulvey, J.M. (eds.) Worldwide Asset and Liability Modeling, pp. 53–61. Cambridge University Press, Cambridge (1998)

[11] Chow, K., Denning, K.C.: On variance and lower partial moment betas: The equivalence of systematic risk measures. Journal of Business Finance and Accounting 21, 231–241 (1994)

[12] Chunhachinda, P., Dandapani, K., Hamid, S., Prakash, A.J.: Portfolio selection and skewness: evidence from international stock market. Journal of Banking and Finance 21, 143–167 (1997)

[13] Corazza, M., Favaretto, D.: On the existence of solutions to the quadratic mixed-integer mean-variance portfolio selection problem. European Journal of Operational Research 176, 1947–1960 (2007)

[14] Dombi, J.: Membership function as an evaluation. Fuzzy Sets and Systems 35, 1–21 (1990)

[15] Elton, E.J., Gruber, M.J.: Modern Portfolio Theory and Investment Analysis. John Wiley & Sons, New York (1995)
[16] Fama, E.: Portfolio analysis in a stable paretian market. Management Science 11, 404–419 (1965)
[17] Fang, S.C., Rajasekera, J.R., Tsao, H.S.J.: Entropy Optimization and Mathematical Programming. Kluwer Academic Publishers, Boston (1997)
[18] Fisher, L., Lorie, J.H.: Some studies of variability of returns on investment in common stock. Journal of Business 43, 99–134 (1970)
[19] Fishman, G.S.: Monte Carlo: Concepts, Algorithms, and Applications. Springer, New York (1996)
[20] Gen, M., Cheng, R.W.: Genetic Algorithms and Engineering Optimization. John Wiley & Sons, New York (2000)
[21] Grootveld, H., Hallerbach, W.: Variance vs downside risk: Is there really that much difference? European Journal of Oprational Research 14, 304–319 (1999)
[22] Hirschberger, M., Qi, Y., Steuer, R.E.: Randomly generating portfolio-selection covariance matrices with specified distributional characteristics. European Journal of Operational Research 177, 1610–1625 (2007)
[23] Hogan, W.W., Warren, J.M.: Computation of the Efficient Boundary in the ES Portfolio Selection Model. Journal of Financial and Quantitative Analysis 7, 1881–1896 (1972)
[24] Holland, J.: Adaptation in Natural and Artificial Systems. University of Michigan Press, Ann Arbor (1975)
[25] Homaifar, G., Graddy, D.B.: Variance and lower partial moment betas as alternative risk measures in cost of capital estimation: A defense of the CAPM beta. Journal of Business Finance and Accounting 17, 677–688 (1990)
[26] Hong, T.P., Chen, J.B.: Finding relevant attributes and membership functions. Fuzzy Sets and Systems 103, 389–404 (1999)
[27] Huang, X.: Fuzzy chance-constrained portfolio selection. Applied Mathematics and Computation 177, 500–507 (2006)
[28] Huang, X.: Credibility-based chance-constrained integer programming models for capital budgeting with fuzzy parameters. Information Sciences 176, 2698–2712 (2006)
[29] Huang, X.: Credibility based fuzzy portfolio selection. In: Proceedings of 2006 IEEE International Conference on Fuzzy Systems, Vancouver, Canada, July, 16-21, pp. 567–571 (2006)
[30] Huang, X.: Chance-constrained programming models for capital budgeting with NPV as fuzzy parameters. Journal of Computational and Applied Mathematics 198, 149–159 (2007)
[31] Huang, X.: Two new models for portfolio selection with stochastic returns taking fuzzy information. European Journal of Operational Research 180, 396–405 (2007)
[32] Huang, X.: Optimal project selection with random fuzzy parameters. International Journal of production economics 106, 513–522 (2007)
[33] Huang, X.: Portfolio selection with fuzzy returns. Journal of Intelligent and Fuzzy Systems 18, 383–390 (2007)
[34] Huang, X.: A new perspective for optimal portfolio selection with random fuzzy returns. Information Sciences 177, 5404–5414 (2007)
[35] Huang, X.: Expected model for portfolio selection with random fuzzy returns. International Journal of General Systems 37, 319–328 (2008)

[36] Huang, X.: Portfolio selection with a new definition of risk. European Journal of Operational Research 186, 351–357 (2008)
[37] Huang, X.: Mean-Semivariance Models for Fuzzy Portfolio Selection. Journal of Computational and Applied Mathematics 217, 1–8 (2008)
[38] Huang, X.: Risk Curve and Fuzzy Portfolio Selection. Computers and Mathematics with Applications 55, 1102–1112 (2008)
[39] Huang, X.: Mean-entropy models for fuzzy portfolio selection. IEEE Transactions on Fuzzy Systems 16, 1096–1101 (2008)
[40] Huang, X.: Mean-variance model for fuzzy capital budgeting. Computers & Industrial Engineering 55, 34–47 (2008)
[41] Huang, X.: A review of credibilistic portfolio selection. Fuzzy Optimization and Decision Making 8, 263–281 (2009)
[42] Huang, X.: Mean-risk model for uncertain portfolio selection, Technical Report
[43] Jana, P., Roy, T.K., Mazumder, S.K.: Multi-objective Mean-variance-skewness model for portfolio optimization. Advanced Modeling and Optimization 9, 181–193 (2007)
[44] Kapur, J.N., Kesavan, H.K.: Entropy Optimization Principles with Applications. Academic Press, New York (1992)
[45] Kapur, J.N.: Maximum Entropy Models in Science and Engineering. Wiley Eastern Limited, New Delhi (1993)
[46] Kaufmann, A.: Introduction to the Theory of Fuzzy Subsets, vol. I. Academic Press, New York (1975)
[47] Kumar, N.V., Ganesh, L.S.: An empirical analysis of the use of the Analytic Hierarchy Process for estimating membership values in a fuzzy set. Fuzzy Sets and Systems 96, 1–16 (1996)
[48] Law, A.M., Kelton, W.D.: Simulation Modelling & Analysis, 2nd edn. McGraw-Hill, New York (1991)
[49] Li, X., Liu, B.: The independence of fuzzy variables with applications. International Journal of Natural Sciences & Technology 1, 95–100 (2006)
[50] Li, X., Liu, B.: Maximum entropy principle for fuzzy variable. International Journal of Uncertainty, Fuzziness & Knowledge-Based Systems 15(suppl. 2), 43–52 (2007)
[51] Li, P., Liu, B.: Entropy and credibility distributions for fuzzy variables. IEEE Transactions on Fuzzy Systems 16, 123–129 (2008)
[52] Liu, B., Iwamura, K.: Chance constrained programming with fuzzy parameters. Fuzzy Sets and Systems 94, 227–237 (1998)
[53] Liu, B.: Dependent-chance programming in fuzzy environments. Fuzzy Sets and Systems 109, 97–106 (2000)
[54] Liu, S.C., Wang, S.Y., Qiu, W.H.: An approach for portfolio selection based on entropy. Systems Engineering: Theory and Practice 18, 245–253 (2000)
[55] Liu, B., Liu, Y.K.: Expected value of fuzzy variable and fuzzy expected value models. IEEE Transactions on Fuzzy Systems 10, 445–450 (2002)
[56] Liu, B.: Theory and Practice of Uncertain Programming. Physica-Verlag, Heidelberg (2002)
[57] Liu, B., Liu, Y.K.: Expected value operator of random fuzzy variable and random fuzzy expected value models. International Journal of Uncertainty, Fuzziness & Knowledge-Based Systems 11, 195–215 (2003)
[58] Liu, B.: Uncertainty Theory: An Introduction to its Axiomatic Foundations. Springer, Berlin (2004)

[59] Liu, B.: A survey of credibility theory. Fuzzy Optimization and Decision Making 5, 387–408 (2006)
[60] Liu, B.: Uncertainty Theory, 2nd edn. Springer, Berlin (2007)
[61] Liu, Y.K., Gao, J.: The independence of fuzzy variables in credibility theory and its applications. International Journal of Uncertainty, Fuzziness & Knowledge-Based Systems 15(suppl. 2), 1–20 (2007)
[62] Liu, B.: A survey of entropy of fuzzy variables. Journal of Uncertain Systems 1, 4–13 (2007)
[63] Liu, B.: Some research problems in uncertainty theory. Journal of Uncertain Systems 3, 3–10 (2009)
[64] Liu, B.: Uncertainty Theory, 3rd edn., http://orsc.edu.cn/liu/ut.pdf
[65] Mao, J.C.T.: Models for capital budgeting: E-V vs E-S. Journal of Financial and Quantitative Analysis 5, 657–675 (1970)
[66] Markowitz, H.: Portfolio selection. Journal of Finance 7, 77–91 (1952)
[67] Markowitz, H.: Portfolio Selection: Efficient Diversification of Investments. Wiley, New York (1959)
[68] Markowitz, H.: Computation of mean-semivariance efficient sets by the critical line algorithm. Annals of Operational Research 45, 307–317 (1993)
[69] Medasani, S., Kim, J., Krishnapuram, R.: An overview of membership function generation techniques for pattern recognition. International Journal of Approximate Reasoning 19, 391–417 (1998)
[70] Medaglia, A.L., Fang, S.C., Nuttle, H.L.W., Wilson, J.R.: An efficient and flexible mechanism for constructing membership functions. European Journal of Operational Research 139, 84–95 (2002)
[71] Michalewicz, Z.: Genetic Algorithms + Data Structures = Evolution Programs, 3rd edn. Springer, New York (1996)
[72] Nawrocki, D.N., Harding, W.H.: State-value weighted entropy as a measure of investment risk. Applied Economics 18, 411–419 (1986)
[73] Pan, Q., Huang, X.: Mean-variance model for international portfolio selection. In: Proceeding of the 2008 IEEE/IFIP International Conference on Embedded and Ubiquitous Computing (EUC 2008), Shanghai, China, December 17-20 (2008)
[74] Parra, M.A., Bilbao, A.B., Uría, M.V.R.: A fuzzy goal programming approach to portfolio selection. European Journal of Operational Research 133, 287–297 (2001)
[75] Peng, Z.X.: Some properties of product uncertain measure, http://orsc.edu.cn/online/081228.pdf
[76] Peng, Z.X.: Uncertainty distribution of functions of uncertain variable, http://orsc.edu.cn/online/090606.pdf
[77] Peng, Z.X., Iwamura, K.: A sufficient and necessary condition of uncertainty distribution, http://orsc.edu.cn/online/090305.pdf
[78] Philippatos, G.C., Wilson, C.J.: Entropy, market risk, and the selection of efficient portfolios. Applied Economics 4, 209–220 (1972)
[79] Philippatos, G.C., Gressis, N.: Conditions of equivalence among E-V, SSD, and E-H portfolio selection criteria: the case for uniform, normal and lognormal distributions. Management Science 21, 617–625 (1975)
[80] Philippe, J.: Value at Risk: The New Benchmak for Controlling Market Risk. Irwin Professional Publishing, Chicago (1996)
[81] Reilly, F.K., Norton, E.A.: Investments, 5th edn. Thomson Learning, New York (1999)

[82] Rom, B.M., Ferguson, K.W.: Post-modern portfolio theory comes of age. Journal of Investing 3, 11–17 (1994)
[83] Roy, A.D.: Safety first and the holding of assets. Econometrics 20, 431–449 (1952)
[84] Rubinstein, R.Y.: Simulation and the Monte Carlo Method. Wiley, New York (1981)
[85] Shannon, C.E.: The Mathematical Theory of Communication. The University of Illinois Press, Urbana (1949)
[86] Simkowitz, M., Beedles, W.: Diversification in a three moment world. Journal of Financial and Quantitative Analysis 13, 927–941 (1978)
[87] Simonelli, M.R.: Indeterminacy in portfolio selection. European Journal of Operational Research 163, 170–176 (2005)
[88] Smimou, K., Bector, C.R., Jacoby, G.: A subjective assessment of approximate probabilities with a portfolio application. Research in International Business and Finance 21, 134–160 (2007)
[89] Tanaka, H., Guo, P., Türksen, B.: Portfolio selection based on fuzzy probabilities and possibility distributions. Fuzzy Sets and Systems 111, 387–397 (2000)
[90] Triantaphyllou, E., Mann, S.H.: An evaluation of the eigenvalue approach for determining the membership values in fuzzy sets. Fuzzy Sets and Systems 35, 295–301 (1990)
[91] Turksen, I.B.: Measurement of membership functions and their acquisition. Fuzzy Sets and Systems 40, 5–38 (1991)
[92] Williams, J.O.: Maximizing the probability of achieving investment goals. Journal of Portfolio Management 46, 77–81 (1997)
[93] Watada, J.: Fuzzy portfolio selection and its applications to decision making. Tatra Mountains Mathematical Publication 13, 219–248 (1997)
[94] Xia, Y., Liu, B., Wang, S., Lai, K.K.: A model for portfolio selection with order of expected returns. Computers & Operations Research 27, 409–422 (2000)
[95] Zadeh, L.A.: Fuzzy sets. Information and Control 8, 338–353 (1965)
[96] Zadeh, L.A.: Fuzzy sets as a basis for a theory of possibility. Fuzzy Sets and Systems 1, 3–28 (1978)
[97] Zhang, W.G., Nie, Z.K.: On admissible efficient portfolio selection problem. Applied Mathematics and Computation 159, 357–371 (2004)

List of Frequently Used Symbols

x	investment proportion, decision variable
ξ	random, fuzzy, uncertain security return
μ, ν	membership functions
μ, e	expected values
σ,	variance value
ϕ	probability density function
Φ, Ψ	probability, uncertainty distributions
$\emptyset$	empty set
Pr	probability measure
$(\Omega, \mathcal{A}, \Pr)$	probability space
Cr	credibility measure
$(\Theta, \mathcal{P}(\Theta), \mathrm{Cr})$	credibility space
$\mathcal{M}$	uncertain measure
$(\Gamma, \mathcal{L}, \mathcal{M})$	uncertainty space
E	expected value operator
V	variance value operator
SV	semivariance value operator
H	entropy operator
α, β	confidence levels
$\alpha(r)$	confidence curve
$R(r)$	risk curve
$\bar{f}$	β-return value
VaRF	Value-at-Risk-in-Fuzziness
VaRU	Value-at-Risk-in-Uncertainty
$\Re$	set of real numbers
$\vee$	maximum operator
$\wedge$	minimum operator
$Eval$	evaluation function in genetic algorithms
GA	genetic algorithm
c	gene
C	chromosome
Err	error function in neural networks

Index